The Resume Catalog:

200 Damn Good Examples

Yana Parker

Author of *The Damn Good Resume Guide*

Ten Speed Press
Berkeley, California

TEN SPEED PRESS
P.O. Box 7123
Berkeley, California 94707

Distributed in Australia by Simon & Schuster Australia, in Canada by Publishers Group West, in New Zealand by Tandem Press, in South Africa by Real Books, and in the United Kingdom and Europe by Airlift Books.

Cover design by Brenton Beck
Interior page composition by Jeff Brandenburg/ImageComp
Illustrations by Ellen Sasaki and Akiko Shurtleff

Library of Congress Cataloging-in-Publication Data

Parker, Yana.
The resume catalog.
Includes index.
ISBN 0-89815-891-5
1. Résumés (Employment) I. Title.
HF5383.P354 1996 650.1'4 87-7057

First printing 1996

Printed in Canada

3 4 5 6 7 8 9 10 — 00 99 98 97

CONTENTS

The Art of Resume Writing iv

Introduction 1

Summarizing Your Experience 2

Making Your Resume "Scanner-Friendly" 2

200 Damn Good Resumes 4

... arranged in 13 broad job categories

Management 6–30

Human Resources 31–48

Administration 49–70

Office Support 71–81

Finance 82–95

Technical & Computers . . 96–108

Education 109–128

Therapy & Social Work 129–141

Health 142–159

Marketing & Promotion 160–179

Sales 180–198

Potpourri 199–255

Youth & Students 256–262

Cover Letters 263

25 Tough Problems 272

... and the resumes that illustrate solutions

Resume Formats; Job Objectives; Skill Areas 273

Lack of Experience; Gaps in Work History; Job Titles 274

Incomplete Education & Credentials, Part-time Jobs; Age Discrimination 275

Ancient experience; Embarrassing Jobs; Confusing Names; Tech Skills . . . 276

Personal Statements; Military Experience; Foreign-born; Consulting 277

Index 278–279

Job Objectives 278

General Index 279

The Art of Resume Writing

Writing a resume is a bit like creating a work of art. There's a hint of poetry to it: given strictly limited space and conditions, you try to say who you are, expressively.

Or perhaps it's a word-sculpture that you keep building, chipping away here, moving this over there, trying out another word or phrase or arrangement—until suddenly it works!

Sometimes it feels like polishing silver, revealing the precious radiance hidden beneath.

Or like washing off a dusty mirror and seeing what you look like from a fresh new perspective.

A resume can be like a sophisticated comic strip: you draw little word sketches of yourself, taking appropriate license with the arrangement of dry historical "facts" to tell a *higher truth*:

... what motivates and moves you
... what work your heart wants to do
... where your hidden or not-so-hidden talents are
... what you've done that makes you feel proud
... what calls forth your passion, competence, and loyalty.

A fine resume is like a flattering snapshot: it captures you at your best, revealing your unguarded beauty.

When it's really good, you have a surprisingly intimate portrait giving the reader an advance clue to your essence. . . telling them what they *really* want to know: *what's special about you.*

Helping someone write a good resume is very gratifying work. —YP

Introduction

Welcome to this second edition of *The Resume Catalog: 200 Damn Good Examples.* This book is a companion and sequel to my popular first book, *The Damn Good Resume Guide*, and contains 200 resumes written in the "damn good" tradition. The *Catalog* is especially useful to job-hunters who learn best by seeing lots of examples of what a well-written resume looks like. Professionals who help *other* people write their resumes will find it equally helpful as a source of ideas on how to structure, write, and graphically design effective resumes.

This collection of resumes represents my work with hundreds of job hunters, and everything here is REAL—real job seekers, real employment histories, real job objectives. Nothing is "made up" although I have changed people's names, addresses, and phone numbers to protect their privacy.

If you look carefully, you'll notice that the job-hunters represented here did NOT usually have "perfect" backgrounds. There were often gaps in their paid employment history, for example—and sometimes the job-hunters had little or no direct experience in the jobs they chose to apply for. These problems had to be resolved or minimized by various strategies shown in the "25 Tough Problems" section.

I mention this because YOU may think YOUR resume has to be "perfect" to be competitive. *Not so.* However, in order to be a *damn good resume,* one that works for both you and a potential employer, your resumes DOES need to:

- Present you at your highest level of competence, playing down any limitations and emphasizing your strengths;
- Focus on a specific job objective and present those of your talents that are most relevant to that objective;
- Link your talents and objectives to the needs of a potential employer.

The sample resumes in this book reflect those goals, and I trust you'll find here some ideas and inspiration to apply to your own resume writing.

Acknowledgments

My heartfelt appreciation goes to the small team who helped put this book together: Patricia Mahoney (planning and file transfer), Ron Weisberg (resume design), Keely Kolmes (planning and project wrap-up), and Valerie Lauer (proof-reading and copying). I am also grateful to the 200 resume clients who allowed their resumes to be shared here—and the job hunters and resume writers who contributed eight of the resumes in this collection. *Thank you.*

—Yana Parker
November 1996

Summarizing Your Experience

You'll notice that a "Summary" or "Profile" appears at the top of most of the resumes in this collection. This feature is a fairly recent innovation in resume writing, and distinguishes a "damn good resume" from, for example, a curriculum vitae, in that a "damn good resume" is most definitely a self-marketing piece of writing.

The Summary is meant to quickly capture the reader's attention and interest—much like a headline on a newspaper article. It should outline the crucial information and inspire the reader to delve deeper for the details.

A good summary, then, could point out:

- The number of years or months of experience in the field.
- Your education, training or certification in that field.
- An accomplishment or recognition that "says it all," if possible.
- Your key skills, talents, or special knowledge related to the target job.
- Something unique about your personal work style or attitude toward the job, that would look appealing to an employer.

Check the effectiveness of your resume by asking yourself:

- Is every item in my Summary *relevant* to my job target?
- Have I *backed up* all the Summary statements through the one-liners in the body of my resume?

Making Your Resume "Scanner-Friendly"

Many large companies now use a computer program such as Resumix to electronically scan applicants' resumes into a computer database. For the company, it's much cheaper and easier to file all those resumes in the computer—because they can later search them electronically for certain key words and quickly identify applicants qualified for a particular job opening. (The computer ignores all the resumes that don't contain the right key words.)

But resume scanning can work for YOU as well as for the employer. You can call the company and find out, *before* submitting your resume, whether they use an electronic resume scanner. If they say "yes" then you need to be sure YOUR resume contains all the "magic words" their computer is programmed to search for.

But what ARE those magic words? Exactly what the computer looks for depends on the job opening, and you can find most of the "magic words" in the company's job description or classified ad for that job.

Take a red pencil and circle all the KEY WORDS that describe the qualifications, experience, skills and any other requirements for the job. *Then, make sure all those KEY WORDS show up on your resume.* It's best to work those KEY WORDS into your "juicy one-liners" that describe your accomplishments.

But just to be SURE you've covered the bases, you could add a paragraph at the bottom of your resume called "KEY WORDS" where you put ALL the key words (and even the *variations* of all the key words) that appear in the job description *and also are true about you and your experience.* In this Key Word paragraph you can apply a policy of "More Is Better," and include all your industry jargon and buzz words. *If in doubt, put it in!* It won't hurt to have too many key words (computers don't get bored), but it *might* hurt to have too few.

Finally, to be sure your resume is truly "scanner-friendly"—meaning the computer has no trouble reading it accurately—make these changes:

In the body of your resume...

- Remove any italics, bold, and underlines.
- Remove any shading, graphics, and decorative lines.
- Remove anything on the *first line* other than your name.
- Use only one "plain vanilla" type font and keep all text at least 10 points in size but no larger than 14 points.

In the key-word section...

- Separate the key words with periods or commas.
- Make sure all the key words are NOUNS*—for example: Purchasing. Raw Materials. Electronic Components. Manager. Amplifier Circuit. AAS Degree in Electronics. Technology. Bookkeeper. General Accounting. Lotus 1-2-3. Drafting. Blueprints. Product Development. OSHA. Training. Freight Operations. HVAC, etc.

* Since your resume may be full of action VERBS (managed, developed, purchased) the NOUNS in your Key Word paragraph may be just what's needed to satisfy the scanner's search criteria.

Still have questions about resume scanning? Get Joyce Lain Kennedy's book, *The Electronic Resume Revolution*, John Wiley & Sons, Inc., 1994, which covers *everything* you need to know about resume scanning. (The guidelines above were adapted from that resource.)

200 Sample Resumes

...arranged in 13 broad Job Categories

NOTE: There is an alphabetical index of all these job titles (objectives) on page 278.

MANAGEMENT, 6–30

7	Alexander Rogers	Drug quality control management
8	Charmaine Duncan	Management, nonprofit organization
9	Erin Irwin	Management, healthcare agency
10	Elizabeth Wool	Executive/admin, community svc
12	Diane Christofaro	Management consultant, healthcare
14	Frances Richardson	HMO administrator
15	Helen Beeson	Asst clinical nursing coordinator
16	Hellmut Dietrich	Administrative director, nonprofit
18	Molly Peterson	Office director for association
19	Janet Crittendon	Office management, interior design
20	Janice Spear	Manager, customer support
21	Judith Brownelle	Director, small biz center
22	Marla Moses	Director/nonprofit agency
24	Michael Hassid	Chief rehab officer, city agency
26	Brenda Gilbert	Management services officer
28	Willard Townsend	Executive director/redevelopment
29	Rita Sasaki	Executive asst, international trade
30	Sandra Hoenig	Food franchise management

HUMAN RESOURCES, 31–48

32	Veronica Silva	Admin analyst/healthcare
33	Tyra Beach	Training specialist, corporate
34	Carol Weitzell	Personnel analyst
35	Marsha Rifenberg	Administrative, human resources
36	Margaret Dwaite	Recruiter/dept forestry
37	Joyce Meyers	Program director, human services
38	Linda Jamieson	Administrative analyst, development
40	Mary Eddy	Human resources/administration
42	Bill Unger	Benefits advocate
44	Anne Hayward	Human resources trainer
46	Mary Newburgh	Analyst, organizational development
48	Stephen Honda	Admin analyst, relocation

ADMINISTRATION/COORDINATION, 49–70

50	Christine Gade	Project supervisor/construction
52	Edith Levenson	Public information officer
54	Deborah Elstad	Program director
56	Christie Keller	Administration, performing arts
57	Anne Fullbright	Events coordinator
58	Lani Simpson	Admin/manufacturing support
59	Ellen Cummings	Project assistant
60	Heather Arnold	Program management
62	Grace Flanders	Fund raiser, development
64	Shirley Krenz	Project management/customer svc
65	Mary Quinlan	Administrative, international trade
66	Nancy Chicago	Assoc director/college admissions
67	Susannah Holt	Volunteer coordinator, humane soc
68	Gelia Thornton	Purchasing/inventory control mgr
69	Roberta Swan	Program management, social svc
70	Rose Ellington	HMO member relations manager

OFFICE/PROGRAM SUPPORT, 71–81

72	Anthony Gabrielle	Customer service, entry
74	Carol Parker	Bookstore clerk
75	Darlene Jacobson	Program assistant
76	Sylvene Piercy	Office support/project coord
77	Estelle Havens	Clerical/bookkeeping, part-time
78	Joy Holland	Administrative assistant
79	Maryanne Hain	Exec asst, international trade
80	Stephen Parker	Mailroom Assistant/Maintenance handyman
81	Stephen Scott	Office support/customer service, entry

FINANCE & ACCOUNTING, 82–95

83	Munana Fehreshta	Accounting, entry level
84	Polly Kelsa	Accounting/financial consulting
85	Suzanne Chew	Auditing, entry level
86	Sonia Morena	Accounts receivable manager
87	Mack Anderson	Management trainee, accounting
88	Andrea Graham	Auditing, entry level
89	Claudia Peterson	Service rep, accounting software
90	David Douglas	Financial analyst
91	Hannah Cortland	Bookkeeper/receptionist
92	Katherine Lawrence	Bookkeeper, full-charge
93	Lorna Burlingame	Financial consultant, investments
94	Lynne Joussart	Business management/financial planning
95	YonSoon Kwang	Junior accountant

TECHNICAL & COMPUTERS, 96–108

97	Susan Pieper	Computer software support
98	Lydia Silvers	Applications engineer, plastics
100	Rochelle Normandy	Computer/PC support services
101	Coronet Gibson	Research services coor, computer firm
102	Norman Whitmore	Technical writing/scientific writing
103	Mark Ebrahimi	Telecommunications supervisor
104	John Bridges	Research associate/biotechnology
105	Larry Elton	Engineering management
106	Patricia Delich	Instructional design/online technologies
108	Michael Hayakawa	Imaging services, director (radiologist)

EDUCATION, 109–128

110	Nicholas Stavrinides	Educational program development
112	Kathleen Webster	Library assistant
113	Tudy Larkspur	Career counseling
114	Charles LaBuz	School counselor, secondary
116	Christiane Lloyd	Instructor, language
117	Fereshteh Ashkani	Childcare or teaching, daycare
118	Sharon Zimmerman	Art teacher
120	Elizabeth Woolsey	Consultant, computer educational prog dev't
122	Mariana Kadish	Translator/interpreter, Spanish
124	Martha Jupiter	Teacher/children's mental health
125	Michelle Olson	Workshop presenter, career counseling
126	Sarah Whitaker	Adult education teacher
128	Sandra Dietz	Substitute teacher

THERAPY & SOCIAL WORK, 129–141

130	Hannah Jenkins	Counselor/addictions
131	Roslyn Marcus	Clinical social worker
132	Patrick Enright	Therapist/social worker
133	Gregory Ackerman	Mental health treatment intern
134	William Ernest	Psychotherapist
135	Carol Weitzell	Program development, elderly
136	Donnette Frost	Chemical dependency intern
138	Eileen Schulman	Social worker
140	Benjamin Farber	Psychotherapist

HEALTH, 142–159

143	Margo Segall	Medical investigator/counselor
144	Julia Millhouse	Manager, geriatric care facility
146	Harriet Bloom	Health educator/nutritionist
148	Harriet Bloom	Project coordinator, special study
150	Robert Adminster	Program director/gerontology
151	Ken Chesak	Body therapist/affiliate
152	Patricia Raines	Stress reduction specialist
154	Richard Jennings	Field supervisor, EMT
156	Pauline Masterson	Director nursing services
158	Carole Loomis	Med/nursing/consultant
160	Tony Foote	Physical therapy aide

MARKETING, 160–179

161	Tracy Hazelton	Market research/analysis
162	Andrea Hughes	PR/marketing
163	Claudia Giselle	Marketing/PR/promotion
164	Cynthia Mayer	Marketing/sales/client services
165	Elizabeth Julian	Publicist/arts organization
166	Fran Morgan	Sales/marketing, PR, promotion
168	Gary Bradley	Marketing, sales management
170	Gary Rosekrans	Accounts executive, ad agency
171	Gerald Davis	Sales & marketing
172	Betsy Emory	Project director, PR officer
173	Jean Bogart	Client services rep, medical center
174	Joanne Simpson	Market research, entry level
175	Joanne Fine	International marketing
176	Elizabeth Leonard	PR/promotion
177	Leslie Bowman	Fashion special events coordinator
178	Linda MacKinnon	Client services rep, health services
179	Sandra Cerrito	Marketing manager, medical products

SALES, 180–198

181	Amy Kurle	Sales/customer service
182	Deborah Richardson	Corporate sales rep, hotel
183	Denise Walters	Sales rep, Kodak
184	Donna Cole	Sales/medical product
185	Ellen Metcalfe	Sales rep/manufacturer's rep
186	Hollis Ann Pope	Buyer/sales, merchandise
187	Jerry Wilcox	Outside sales rep
188	Jerry Parkhurst	Merchandising display
189	Tricia Baker	Sales management, consumer services
190	Judy Rogers	Sales/marketing
191	Linda Mowry	Sales rep/account executive
192	Mark Fleetwood	Sales rep, electronics
193	Melinda Sailor	Jewelry sales/customer service
194	Noreen MacLaughlin	Sales
195	Pamela Swiss	Outside sales rep, fashion
196	Sherrie Valencia	Sales/marketing
197	Vreny Zurich	Small store manager, shoes
198	Eleanor Kennedy	Sales rep

POTPOURRI, 199–256

• Environmental

200	Barbara McClosky	Info specialist/environmental
201	Brian Rebar	Environmental technician

• Product Development

202	Emily George	Clothing designer/illustrator
204	Joyce Stroebech	Production assistant, clothing
205	Rebecca Vaness	Production assistant, clothing
206	Wendy Gillroy	Product developer/client services rep

• Film, Radio, TV

207	Katherine Brunswick	Film production assistant
208	Lynn Sheffield	Radio or TV programming/announcing
209	Martha Azcona	TV programming production
210	Rebecca Newburg	Film production manager/assistant
211	Roger Lancaster	Entry level, TV news grip

• Property Management

212	Adrienne Mendoza	Property management, entry level
213	Mark Killorin	Property management trainee
214	Richard Flores	Property management, Wells Fargo

• Graphic Design

215	Amy Buchannon	Graphic designer assistant
216	Barbara Monet	Graphic production
217	Gregory Byron	Entry level, multi-image production
218	Lynne Charney	Design/production, publications
219	Vickie Wan	Design/multi-media

• Legal

220	Ann Voorhees	Staff attorney
222	Michael Oliver	Law office clerk, research assistant
223	Randolph Strough	Paralegal
224	Andrew Thompson	Legal assistant

• Editing/Writing

225	Richard Griffon	Freelance editing, proofreading
226	George Amundsen	Editorial assistant w/Chevron
227	Rebecca Bridges	Editorial assistant/publishing
228	Mary Moriarity	Freelance book editing
230	Carolyn Clarke	Editorial assistant, book publishing

• Real Estate Appraisal

231	Linda Durkee	Real estate appraiser, entry
232	Margaret Lester	Real estate appraiser trainee
233	Dennis Brinkley	Real estate appraisal trainee

• Religion

234	Susan Holmes	Pastoral minister
236	Sister Mary Jones	Pastoral associate

• One-of-a-Kind

238	Jo Anne Burgess	Union rep/business agent
239	Bradley French	Writer/photographer/editorial assistant
240	Loren Greene	Summarizer, Barron's legal services
241	Michael Wong	Community or gov't relations rep
242	Richard Jennings	Fire fighter, entry level
244	Robert Lawton	Service writer/auto manufacturer
245	Stephen Honda	Real estate analyst/researcher
246	Donald Raulings	Private investigator
248	Teresa Fernandez	Wardrobe assistant, movie/TV
250	Kenneth Richards	Warehouseman
251	Carlene Doonan	Apprentice baker
252	Dawn Ellsworth	Gallery assistant/sales
253	Hellmut Dietrich	Freight handling, import/export
254	Dolores Walker	Commercial leasing, agent
255	Denise Francis	Activity director, cruise staff

YOUTH & STUDENTS, 256–262

257	Matthew Kurle (age 8)	Nothing listed yet
258	Julia Smith (age 14)	Baby-sitting /odd jobs
259	Charlie Knych	After-school jobs (high-school student)
260	Karen Schmidt	Marketing trainee (college student)
261	Valerie Lauer	Entry, office services (college-bound)
262	Rachael Hennesey	Internship, public law (college student)

Management Resumes

Alexander Rogers	Drug quality control management	7
Charmaine Duncan	Management, nonprofit organization	8
Erin Irwin	Management, health care agency	9
Elizabeth Wool	Executive, community service	10
Diane Christofaro	Management consultant, health care	12
Frances Richardson	HMO administrator	14
Helen Beeson	Assistant clinical nursing coordinator	15
Hellmut Dietrich	Administrative director, nonprofit	16
Molly Peterson	Office director, association	18
Janet Crittendon	Office manager, interior design	19
Janice Spear	Manager, customer support	20
Judith Brownelle	Director, small business center	21
Marla Moses	Director, nonprofit agency	22
Michael Hassid	Chief rehabilitation officer, city agency	24
Brenda Gilbert	Management services officer	26
Willard Townsend	Executive director, redevelopment	28
Rita Sasaki	Executive assistant, international trade	29
Sandra Hoenig	Food franchise management	30

ALEXANDER ROGERS

2933 Broderick St.,
San Francisco, CA 94115
(415) 998-3654

Objective: Position in Quality Control Management with the Food, Drug, and Cosmetics industry.

SUMMARY

- 18-year track record in managing a total quality control system.
- Expert troubleshooter in both manufacturing and packaging.
- Thorough working knowledge of the food / drug / cosmetic industry.
- Results oriented; confident in making on-the-spot decisions.
- Industry reputation for professionalism and competence.

PROFESSIONAL EXPERIENCE

Quality Control Management

- Successfully located and qualified the best contract manufacturers in the industry, consistently meeting all regulatory and corporate specifications.
- Designed and implemented effective quality / cost control systems:

  Contract manufacturing
  - Placed over 100 products with contract manufacturer, achieving a trouble-free transition from in-house manufacturing.
  - Maintained a product complaint level below the industry norm.

  In-house manufacturing

  Managed a $1.3 million departmental budget:
  - Increased departmental productivity by 65%;
  - Decreased product rejection rate by 30%.

Troubleshooting

- Restored efficiency of Component Inspection Dept. at Max Factor in 3 months:
  - Retrained staff;
  - Automated inspection procedures;
  - Rewrote procedures;
  - Introduced time management.

Industry Expertise/Product Knowledge

- Wrote procedure manual detailing the requirements for qualifying contract manufacturing with Shaklee.
- Wrote procedure manuals for Quality Control program at Max Factor & Co.
- Active in promoting industry technical education:
  - Spoke on qualification of contract manufacturers, at both ASQC and SCC;
  - Served on national Executive Committee of ASQC, FD&C section;
  - Selected Publicity Chairman, 1995, ASQC, FD&C annual West Coast seminar.

EMPLOYMENT HISTORY

1987–present	**Quality Assurance Manager**	SHAKLEE, San Francisco Headquarters
1986–87	**Quality Assurance Manager**	SHAKLEE, Hayward
1984–86	**Quality Control Supervisor**	MAX FACTOR & CO., Los Angeles
1983–84	**Quality Control Manager**	DRACKETT INC., Los Angeles
1979–83	**Quality Control Supervisor**	CARTER-WALLACE INC., Los Angeles

EDUCATION

B.S., Business Administration – UNIVERSITY OF REDLANDS, CA

Charmaine's Army experience is interpreted in civilian language.

CHARMAINE DUNCAN

361 - 25th Avenue • San Francisco, CA 94121 • (415) 339-4421

OBJECTIVE: Management position in a nonprofit organization.

SUMMARY

- Six years' successful supervisory and management experience.
- Resourceful and self-confident; get the job done, and do it well.
- Strong interpersonal and communication skills.
- Extensive experience in design and implementation of training programs.
- Remain calm and work well under demanding conditions.

PROFESSIONAL EXPERIENCE

Management

- Established training programs:
  - Wrote and published yearly training calendars;
  - Located and scheduled instructors for weekly classes in military skills;
  - Coordinated training for personnel dispersed over 1.7 million miles;
  - Forecast and managed annual training budget while maximizing formal schooling.
- Planned, coordinated and conducted management conferences attended by officers from all over the country.
- Managed a small convenience store, overseeing orders and deliveries, and supervising three employees.

Supervision

- Exercised total supervisory responsibility for a unit of 22 security personnel:
  - Organized the unit into teams and designated the first-line supervisors;
  - Established an evaluation program, utilizing quarterly written reports;
  - Directed field operations: planning, budgeting, travel and housing arrangements.
- Successfully planned, organized and led field training missions in a "real-world" environment, ensuring adequate provisions and operational equipment, and providing direction and supervision for 30-40 trainees for up to three weeks.

Written & Oral Communication

- Made presentations to officials to get approval and funding for operations.
- Wrote and presented info briefings to peer officers to maximize use of resources.
- Effectively counseled team leaders and supervisors.
- Authored award recommendations for subordinates that consistently won approval.
- Compiled and edited comprehensive activity reports from subordinate units for national publication.

WORK HISTORY

1995–present	***Administrative Asst.***	ALUMNAE RESOURCES Career Ctr., SF
1986–94	***Chief, Tech. Security Branch***	U.S. ARMY, SF
	Battalion Plans & Training Officer	" "
	Chief, Collection Mgt. Section	U.S. ARMY, Germany
	Platoon Leader	" "
1984–86	***Student***	
1982–83	***Manager***	RICHLAND CORP., Atlanta, GA
1981–82	***Admin. Asst./Bookkeeper***	STATE BANK & TRUST, Atlanta, GA

EDUCATION

B.A., German – Georgia State University, Atlanta, GA

ERIN IRWIN
724 Filbert Street, Apt. 5
San Francisco, CA 94123
(415) 688-7816

Objective: Management position with a health care agency.

SUMMARY

- Strong combination of management and clinical experience.
- Expertise in health care systems through 17 years in the field.
- Natural talent for public relations and marketing.
- Successfully developed and controlled a $1.3 million budget.
- Supervised a staff of 32 professionals and support personnel.

RELEVANT EXPERIENCE

Management & Supervision
- Developed $1 million budget covering unit's personnel, supplies and equipment.
- Audited and analyzed unit-level needs for supplies and equipment.
- Interviewed and hired personnel, both clinical and support staff.
- Facilitated staff planning meetings, promoting high level of goal achievement.
- Developed effectiveness and efficiency of clinical staff at St. Joan's nursing unit:
  - Upgraded quality of training and orientation of both new and old staff;
  - Instituted a unit-based management group, improving communication;
  - Evaluated individual work performance and advised on career development.

Public Relations/Community Liaison
- Followed up on a day-to-day basis with patients on adequacy of their care.
- Identified problem issues and effectively arranged for best resolution possible.
- Coordinated discharge planning with outside health care agencies.

Counseling/Advocacy
- Counseled patients in crisis and terminal care situations; counseled staff on both personal and work issues.
- Served as advocate for clients, cutting through red tape to assure adequacy of patient care and access to health care information.
- Successfully mediated conflicts among staff members; between staff and patients; and between patients and family.

EMPLOYMENT HISTORY

1986–96	**Head Nurse**	ST. JOAN'S HOSPITAL	Cardio-Vascular Surgical Unit
1983–86	**Senior Staff Nurse**	" "	Surgical Intensive Care Unit
1981–83	**Senior Staff Nurse**	" "	General Surgical Unit
1979–81	**Staff Nurse**	" "	General Surgical Unit

EDUCATION

M.S., Human Resources & Organizational Development – USF., San Francisco
B.S.N. – USF., San Francisco; Licensed California RN

ELIZABETH WOOL

29 Ramsgate Lane
Pleasant Hill CA 94523
(510) 555-2136

Objective: Executive/administrative position with a broad-based community services agency.

SUMMARY

- Over 10 years' administrative and management experience.
- Practical working knowledge of every level of agency operations.
- Solid theoretical training in management.
- Proven record of innovative and effective staff development.
- Strong commitment, vision and leadership.

WORK HISTORY

1994–present ***Community Relations Coordinator*** AMERICAN RED CROSS, Richfield, CA

- Served, by invitation of Community Council, as representative entrusted to negotiate with the University for space and financial support of child care program.

1993–94 ***Program Specialist*** (consultant) PARENTAL STRESS CENTER, Pittsburgh, PA

- Designed and conducted an innovative 6-week intensive training program for child care personnel, with unique opportunities for in-service training leading to college enrollment.
- Set up improved filing systems for statistical data and program record keeping.

1989–92 *Student, part-time* UNIVERSITY OF PITTSBURGH

– Concurrent part-time work: Insurance & Sales; Census Enumerator; Researcher

1985–89 ***Training Specialist*** PITTSBURGH CHILD GUIDANCE CTR. Pittsburgh, PA

- Conducted teacher training sessions for private and public school teachers:
  - – Implemented needs assessment of existing teaching staff;
  - – Developed training program specifically for identifying and managing (in classroom) pre-school and school-age children with emotional and learning disabilities;
  - – Trained teachers in special testing and tutoring skills for use with learning-disabled children.

– Continued –

1983–85 ***Director***, "Home Start" Project — ACTION INC., Gloucester, MA
National Research & Demonstration Proj.

- Provided highly effective management for a 3-year national demonstration research project:
  - Established cooperative spirit by developing channels for exchange of expertise and resources;
  - Built highly productive work teams, introducing many opportunities for staff input;
  - Convinced skeptical manager to support the program, through effective negotiations and follow-through.
- Built up an effective advocacy program for this model federally-funded program:
  - Successfully involved all staff members in developing a comprehensive network of community services and contacts;
  - Developed a slide show presentation explaining the program to other agencies;
  - Gained the support of other organizations to provide client services needed to achieve program goals.

1978–82 ***Director*** — HASSETT DAY CARE CENTER, Boston, MA

- Secured funding for three child development programs, writing successful grant proposals.

EDUCATION

M.E., Education – Northeastern University, Boston, MA
B.A., Government – Boston University, Boston, MA
Graduate study in Public Administration – University of Pittsburgh, PA
Training in Human Resource Development – Certificate Program, UC Berkeley Extension

DIANE H. CHRISTOFARO

Health Care Management Consultant

P.O. Box 45678
Uniondale, PA 18711
(717) 234-5678

SUMMARY

- 15 years' experience in health care management.
- Recognized as an authority on managed care programs.
- Professional credentials in nursing and health education.
- Successfully administered joint venture for Blue Cross.
- Developed innovative programs for major HMO, incorporating trends on the leading edge of health care delivery.

PROFESSIONAL ACCOMPLISHMENTS

HEALTH MANAGEMENT JOINT VENTURE

Managed joint venture, Eastern Managed Care Inc., from initial start-up through full implementation as a successful and self-sustaining corporation:

- Defined corporate goals and objectives.
- Set up the operational systems, with enhancements as needed.
- Authored operational and program policies and procedures.
- Recruited and trained management staff, field staff and support staff; built staff to a national network of 40 professionals.
- Developed budget ($.5 million first-year, $1 million second-year) incorporating long-term planning and marketing strategies.
- Wrote proposals for submission to state and federal agencies, employers and HMOs.
- Negotiated contracts with employers, HMOs and insurance companies.
- Maintained high level of program effectiveness, constantly evaluating performance against industry standards.
- As recognized authority in the field of Health Benefits Management, addressed the National BC/BS Conference on "Psychiatric and Substance Abuse Case Management."

HMO CONTRACT MANAGEMENT

Successfully obtained contract with a major HMO, being the only company that could provide the needed expertise in comprehensive managed care.

- Designed utilization review standards and protocols.
- Incorporated a unique component, "Case Management," to ensure the most medically appropriate and cost-effective care and treatment.
- Upgraded the reporting procedures, providing accurate measures of program success.

CAPITAL BLUE CROSS PROGRAM DEVELOPMENT

Developed a department within Plan to provide case management services to external accounts.

- Designed customized software program enabling the Department to efficiently obtain, store and retrieve critical claimant information, essential for quality case management.

Continued on page 2

"Health Care Management Consultant" is used in place of "Objective: . . ." because she's looking for an independent consultant relationship rather than employee status.

CAPITAL BLUE CROSS PROGRAM DEVELOPMENT, continued

- Wrote manuals for Operations, Case Management, and Marketing Training for field staff, account reps, and administrators.
- Advised management on program pricing, advertising and marketing strategies.

EMPLOYMENT HISTORY

1996–present	**Executive Director**	EASTERN MANAGED CARE INC, Wilkes Barre, PA
1995–present	**Vice-President**	MOORE HEALTH MANAGEMENT INC., Managed Care Services, Oneonta, NY
1994–95	**Owner/Principal**	INDUSTRIAL HEALTH DESIGNS, health care management consulting, New York State
1993–94	**Rehabilitation Consultant**	NATLSCO Rehab Management Inc., NY and PA
1991–93	**Rehabilitation Specialist**	INTERNATIONAL REHAB ASSOC., Missoula, MT
1991–92	**Health Educator**	FIVE VALLEYS HEALTH CARE, Missoula, MT
1990–91	**Director, Medical Services**	ONEONTA, NY JOB CORPS CENTER
1989–90	**Interim Director**	STUDENT HEALTH SERVICES, Delhi, NY Delhi Agricultural & Technical Institute
1982–89	**Critical Care Nurse/Supv.**	Hospitals in New York and Colorado

EDUCATION & CREDENTIALS

M.A., Social Sciences – STATE UNIVERSITY OF NEW YORK, BINGHAMTON
B.S., Education, STATE UNIVERSITY OF NEW YORK, ONEONTA
A.A.S., Nursing (honors), ORANGE CO. COMMUNITY COLLEGE, Middletown, NY
Psychology, UNIVERSITY OF COLORADO, Boulder

CERTIFICATIONS & LICENSES
Licensed Registered Professional Nurse – New York
Certified Facilitator, National Center of Health Promotion

PROFESSIONAL DEVELOPMENT

1995 National Medical Case Management Seminar, INTRACORP, Dallas TX
1995 POCONO Conference, "The Future of Health Care," Wilkes Barre PA
1996 Annual Risk Management and Employee Benefits Conference, Las Vegas, NV

PROFESSIONAL AFFILIATIONS

Institute for the Advancement of Health
National Center of Health Promotion
National Association of Executive Females
National Rehabilitation Association
American Society for Training and Development
Association of Rehabilitation Nurses

FRANCES RICHARDSON

333 - 23rd Avenue•San Francisco, CA 94121•(415) 489-9000 work•(415) 233-4141 home

Objective: **Management position with a hospital or HMO,**
applying experience in environmental services, safety, education,
marketing, development, guest relations, and/or physician relations.

SUMMARY OF QUALIFICATIONS

- Ten years' professional experience, with a degree in Health Services Administration.
- Born leader; inspiring others to work at their highest level.
- Proven management skills and record of accomplishment.
- Highly creative and innovative, not afraid to take risks.

PROFESSIONAL ACCOMPLISHMENTS

Environmental Services • Safety • Education

- As Director of Housekeeping & Linen Departments , reduced budget and staff costs by 30%, while maintaining highest quality service.
- As Safety Director, managed Workers Compensation Program and successfully reduced loss ratio from 35% to 11% in one year, resulting in major savings for the hospital.
- Served as expert advisor for a videotape on hospital disaster planning, produced for nationwide distribution by Hospital Satellite Network.

Guest Relations • Physician Relations

- Researched and selected Guest Relations Program, a major hospital investment:
  – Selected most effective potential staff trainers;
  – Co-designed detailed implementation plan.
- Introduced highly successful programs, enhancing rapport between hospital and physicians, for example:
  – Practice Enhancement Program, which attracted personal involvement of the physicians;
  – Seminar on use of CO2 Laser, providing Continuing Education credit for physicians and increasing use of the hospital's equipment and surgery revenue.

Marketing • Special Events Coordination

- Developed innovative marketing concepts and strategies, assisting Director of Marketing in: ...new program identification, development and promotion; ...research ...advertising campaigns ...media liaison ...ad agency selection.
- Organized all aspects of successful major fund-raising events, benefiting:
  – Coming Home Hospice, a subsidiary of VNA (a hospice for AIDS patients);
  – St. Vincent de Paul Society shelters for the homeless.

EMPLOYMENT HISTORY

1995–present	***Director, Customer Relations***	HUNTINGTON HOSPITAL, San Francisco
1993–95	***Environmental Services Mgr.***	HUNTINGTON HOSPITAL, San Francisco
1991–93	***Director, Housekeeping & Linen***	HUNTINGTON HOSPITAL, San Francisco
1987–91	***Environmental Services Asst.***	HUNTINGTON HOSPITAL, San Francisco

EDUCATION

B.A., Health Services Administration, with honors – ST. MARY'S COLLEGE, Moraga - 1993

Helen lists each different position with the same employer, showing advancement within her field.

HELEN BEESON, R.N.

1722 Nelson Blvd. ♦ Oakland, CA 94611 ♦ (510) 801-3277

OBJECTIVE: **Position as Assistant Clinical Nursing Coordinator, Intensive Care Nursery at Stanford University Hospital.**

SUMMARY

- Thorough understanding of the nursing process, both in theory and in practice.
- Highly skilled in effective leadership techniques.
- Able to balance the needs of the staff with the priorities of the institution.
- Solid experience in staff orientation, scheduling, evaluation and development.
- Night Shift Coordinator for 3 years at Mount Zion Hospital and 1 year at Children's Hospital, supervising up to 45 nurses.

PROFESSIONAL EXPERIENCE

1987–present MOUNT ZION HOSPITAL Intensive Care Nursery; Pediatrics

Nurse Manager, 1995–present
Interim Manager (6 mos., 1993)
Liaison Nurse, 1989–90
Staff Nurse II, III, 1993–94
Clinical Coordinator, 1990–93
Staff Nurse II, 1987–88

Orientation

- Coordinated and developed programs:
  – Education program for professional organization's annual symposium;
  – Outreach education for Level I and Level II nurseries;
  – Orientation program and in-service programs on a variety of subjects:
  ...transport of sick infants; ...nursing interventions with families of sick infants;
  ...contemporary teaching/learning theories; ...resuscitation of the newborn.
- Oriented ICN staff and outreach nurses from other hospitals.
- Authored numerous policies and procedures for nursing care.

Counseling & Development

- Developed a highly effective strategy for working with "difficult employees":
  – Determined whether there is potential for positive change;
  – Identified and presented the problem to the employee;
  – Clarified and reached agreement about behavioral expectations;
  – Strongly supported the employee with my confidence in their progress;
  – Developed a plan, including time frame for re-evaluation of progress;
  – Carefully documented the steps taken and the results.

Staff Supervision

- Currently supervise a staff of 80 intensive care nursery nurses and 12 pediatric nurses, on a 24-hour basis.
- Scheduled and staffed intensive care nursery and pediatric units.

1983–84 MOUNT ZION HOSPITAL Intensive Care Nursery

Staff Nurse II, III

EDUCATION & TRAINING

A.A., Nursing, with honors – MERRITT COLLEGE, Oakland
Specialized training: Leadership; Problem Solving, Coaching & Developing, Group Dynamics
A.A., Mathematics & Science, with honors – MERRITT COLLEGE, Oakland

HELLMUT DIETRICH
333 - 65th Street
Oakland, CA 94609
(510) 699-4742

Objective: **Position as Administrative Director with a nonprofit social service agency.**

SUMMARY

- Experience and training in basic office systems.
- Skilled supervisor; able to motivate and handle conflict.
- Competent in programming and operating computers.
- Creative and flexible in organizing and planning.
- Diplomatic and effective in negotiations.

RELEVANT EXPERIENCE & SKILLS

ADMINISTRATION

- Served on Board of Directors for a nonprofit organization serving parents:
  - Wrote successful grant proposals;
  - Planned annual budget;
  - Negotiated contracts for property rentals, insurance, apprenticeships;
  - Interviewed and hired staff, as member of personnel team;
  - Interpreted personnel guidelines, resolved staff conflicts, negotiated wages.
- Negotiated with speakers and artists (singers, writers, parliamentary figures, representatives of minority groups) and made detailed arrangements.

SUPERVISION

- Supervised volunteer teachers and freelance tutors at a drop-in community center for low-income youth.
- Supervised warehouse crew in the shipping and handling of merchandise, as warehouse superintendent at a large freight company in Munich:
  - Initiated improvements in teamwork efficiency by clarifying areas of responsibility and promoting better communication;
  - Developed and updated time schedules and delivery tours for truckers.

PROJECT COORDINATION

- Organized public educational/entertainment events, involving arrangements for location, insurance, contracts, security, advertising and promotion.
- Successfully coordinated community projects involving previously conflicting groups, overcoming hostility and mistrust by identifying common interests and goals.

COMPUTER SKILLS

- Familiar with use of personal computers and printers:
  - Customized commercial computer programs to meet unique agency needs;
  - Wrote complex programs using direct disk access to retrieve lost data.

– Continued –

HELLMUT DIETRICH
Page two

EMPLOYMENT HISTORY

1996–present		Travel and relocation to the Bay Area
1993–95	***Student***	UNIVERSITY OF TÜBINGEN, Germany
1989–93	***Display Builder***	S&E STARK PROMOTIONAL DISPLAY CO., Ostfildern, Germany
1986–88	***Teacher***	Jr. High School, Elementary School; Osfildern, Germany
1983–85	***Teaching Asst.***	PÄDAGOGISCHE HOCHSCHULE COLLEGE, Esslingen, Germany
1982–86	***Student***	" "
1981	***Social Worker***	City of Schwäbisch Hall, Germany
1980	***Warehouse Supv.***	ANTON GLATZ FREIGHT CO., Munich, Germany

VOLUNTEER WORK

1988–90	***Program Organizer***	CITIZENS AGAINST RACISM, Esslingen, Germany
1990–94	***Member, Board of Directors***	PARENT & CHILD SERVICES, Tübingen, Germany

EDUCATION & TRAINING

B.A., Social Work (German equivalent) – University of Tübingen, Germany
Teaching Credential (German equivalent) – Pädagogische Hochschule College
Classes in: Bookkeeping, correspondence, transportation, business and math.

> Hellmut makes sure that all his German work history is clearly described, adding a word or two where needed to explain the employer's business. In the Education section, he points out the German equivalent of US credentials. (See Hellmut's other resume on page 253.)

"Home management, travel and study" account for a major break in Molly's work history.

MOLLY PETERSON

800 Marietta Drive • Los Angeles, CA • (202) 141-1119

Objective: Regional office director of a national preservation organization.

PROFILE

- Lifelong interest in and commitment to architecture and preservation.
- Four years' leadership experience as president and Chairman of the Board of Los Angeles Architectural Preservation (LAAP).
- Fundraising experience during six years' chairing development committee.
- Recent completion of 18 months of graduate work in nonprofit management.
- Competent in communicating the value of historic preservation.
- Recruited presenters and trainers for 100 programs.

PROFESSIONAL EXPERIENCE

Public Relations, Communication, Media

- Spoke as expert in preservation issues and the history and architecture of Los Angeles:
  - Conducted walking tours of residential and commercial areas for past 8 years;
  - Made one-to-one and small group presentations to major donors and corporate leaders.
- Conducted demonstrations for audiences of 100–1000 people throughout the U.S.

Fundraising/Development

- Successfully identified and cultivated major donors & significantly increased contributions:
  - Brought people onto the board who themselves were excellent fundraisers and givers;
  - Increased participation, matching contributors with projects aligned with their interests.
- Conceived, organized, and implemented numerous successful fundraising events:
  - Doubled the profit of The Urban Career Center's annual luncheon, from $12,000 to $24,000;
  - Netted LAAP $15,000 from a building opening and $35,000 from a restaurant opening;
  - Increased, from $30,000/evening to $65,000, the proceeds of LAAP's annual Summer Ball.

Leadership, Management, Administration

- Monitored organizational budgets of $400,000 - $600,000 monthly for eight years.
- Served on the Board of Directors of LAAP and The Urban Career Center:
  - Developed program and personnel policies, and organizational goals and budgets;
- Directed interviews for executive and development directors, and new board members.

WORK HISTORY

Current	**Graduate student**	University of California, Los Angeles
1993–96	**Member, Executive Committee**	Los Angeles Architectural Preservation
" "	**Vice President, Board of Directors**	The Urban Career Center, SF
1988–92	**Chair and President of Board**	Los Angeles Architectural Preservation
1975–87	Home management, travel, study	
1972–74	**Teacher**	Los Angeles public schools

EDUCATION

M.P.A. graduate studies in Nonprofit Management,
University of California, Los Angeles

B.S. – Iowa State University, Ames, Iowa

This resume can easily be made "generic" to fit any other office management setting just by replacing the "Summary" and dropping the last line of the resume.

JANET CRITTENDON
4967 Neilson Blvd.
Albany, CA 94706
(510) 666-3115

Objective: Supervisory or office management position in an interior design firm.

SUMMARY OF QUALIFICATIONS

- Over six years' experience in management and supervision.
- Highly effective in promoting a positive, productive work environment.
- Reputation for excellence and high quality service to clients.
- Good eye for detail; well organized, skilled in setting priorities.
- Lifelong interest in interior design; classwork in space planning & color.

EMPLOYMENT HISTORY

1989–present **Manager**, Plant & Office — ALBANY SURGICAL SHOES, INC., Albany, CA
Manufacturer of shoes and podiatry supplies

- Delegated responsibilities among production foremen, collaborating on daily production priorities and maintaining a smooth flow of operations.
- Purchased manufacturing materials and office supplies, considering production schedules, vendor delivery schedules, and cost-effectiveness.
- Successfully built a cooperative work team and promoted productive environment by:
  – Delegating jobs in accordance with employees' skills and abilities;
  – Treating employees with respect; maintaining a sense of humor;
  – Welcoming constructive criticism and input on production improvements;
  – Supporting morale by posting letters of appreciation from customers.

1987–89 **Office Manager** — ALBANY SURGICAL SHOES, INC.

- Designed office procedures; developed company policy on handling customer complaints.
- Organized and improved filing system for contracted accounts, resulting in substantial time saving for handling government files and mandatory quarterly report.
- Oversaw all facets of record keeping including payroll, insurance, banking, accounts receivable, accounts payable, general ledger.

1983–87 **Office Asst./Customer Svc.** — ALBANY SURGICAL SHOES, INC.

- Developed innovative customer services, building a reputation for our company as a leader in the industry, with unique same-day service & products not available elsewhere.
- Trained employees in developing and maintaining good customer relations, and how to effectively negotiate and resolve customer service problems.

1981–83 **Assembler/Inspector** — ALBANY SURGICAL SHOES, INC.

EDUCATION

SAN FRANCISCO CITY COLLEGE
Business Management, Bookkeeping, Accounting, 1983–85
Interior Design, 1995–96

JANICE SPEAR
119 Hemlock Road • San Francisco, CA 94115
(415) 329-6790

OBJECTIVE Manager of customer support, training, or sales.

SUMMARY
- Ten years' successful experience as a manager.
- Highly effective in motivating and supervising employees.
- Well organized; strong in planning and implementing programs.

EXPERIENCE

PR/Marketing/Sales
- Developed an improved marketing plan:
  - Assembled a photographic presentation of services available;
  - Changed procedure for membership sales, to include facility tour and active closing;
  - Initiated a program of follow-up calls to guests to encourage membership.
- Stimulated direct involvement in social and athletic activities by communicating the excitement and fun of the event.
- Persuaded large corporations and businesses to sponsor events.

Program Development/Coordination
- Selected social and athletic activities to be offered; coordinated implementation, promoted and registered participants.
  - Planned, coordinated and promoted dozens of racquetball tournament events (charity, celebrity, professional).

Supervision & Training
- Recruited, supervised and evaluated 15 employees.
- Authored employees' training manual, describing the facility, all operations, and detailed procedures of the athletic club.
- Maximized staff effectiveness and morale by assigning tasks related to their interests and promoting teamwork over competitiveness.

Management
- Oversaw all aspects of operating an 8-court racquetball club, including full financial accounting of $450,000 annual revenues.
- Forecast and developed $200,000 annual operating budget for $20 million facility, covering: salaries, monthly guest fees, rentals, locker room supplies, uniforms, locks & keys.
- Cut costs 20% by more cost-effective purchasing of supplies.

EMPLOYMENT HISTORY

1995–present	*Hostess*	PRINCESS CRUISE, San Francisco
1992–94	*Operations Supv.*	THE TEXAS CLUB athletic club, Houston
1988–92	*Manager*	WOODLAKE RACQUETBALL CLUB, Houston
1985–88	*Asst. Manager*	COURT SPORTS raquetball club, Houston

EDUCATION B.S., Elementary Education, University of Texas at Austin
Teaching Certificate, Special Education

JUDITH BROWNELLE

8686 Shellmound
Emeryville, CA 94608
(510) 644-2989 (home); (510) 421-8744 (office)

Objective: Position as Director, East Bay Small Business Development Center.

PROFILE

- Strong background combining business, liberal arts and community experience.
- Familiar with Bay Area business, government, education and non-profits.
- Extensive experience in writing and reviewing proposals.
- Work effectively both as team member and independently.
- Enthusiastic, sharp, and well organized.

PROFESSIONAL EXPERIENCE

Program Planning

- Designed and directed a highly successful youth volunteer program for the Peralta Community College District, later chosen as model for other programs nationwide:
  - Persuaded school district to provide critical financial and staffing support;
  - Coordinated joint sponsorship of City, Volunteer Bureau, and community agencies and businesses.
- Reconciled loan payment records between servicing company and 100 lending institutions.

Communications & PR

- Produced wide range of business and programming communications:
  - Edited Teletrends, international newsletter on telecommunications trends in higher education;
  - Wrote program proposals, summaries, evaluations and recommendations;
  - Authored business plans, annual reports and status reports.
- Chaired community meetings, local/regional conferences, planning meetings of college development staff, and professional association meetings.
- Trained faculty and staff in proposal writing and program planning techniques.
- Addressed groups of educators, community and business leaders and students.
- Promoted agency programs through networking and presentations at major conferences.

Research / Analysis / Evaluation

- Reviewed demographic data and labor market projections to establish program directions.
- Developed theoretical knowledge of financial analysis, budgeting, forecasting, statistics and research methods and strategic planning, through MBA case studies.
- Designed and monitored project budgets for community and educational projects.

EMPLOYMENT HISTORY

1994–present	***Resource Developm't Specialist***	PERALTA COMM. COLLEGE DIST., Oakland
1989–93	***Staff Asst./Planning & Dev't.***	" " "
1986–89	***Project Director***	VOLUNTEER CENTER OF ALAMEDA CO.
1985–86	***Program Coord. (VISTA Vol.)***	VOLUNTEER CENTER OF ALAMEDA CO.
1980–84	***Administrative Assistant***	FIDELITY MUTUAL LIFE INS. CO., Philadelphia

EDUCATION

MBA, Executive Program – St. Mary's College, Moraga, CA, 1994
BA, Geography / Urban Studies – University of Pennsylvania, Millersville, 1983

MARLA MOSES

78 Westwind Street • Walnut, CA 95448
(707) 323-7804 or **(707) 823-2929**

Objective: Position as director/associate director with a public or nonprofit agency.

SUMMARY OF QUALIFICATIONS

- 6 years' successful experience in program management and development.
- Resourceful in solving problems and maximizing resources.
- Effective in promoting a positive, productive work environment.
- Talent for balancing long-range vision with attention to detail.
- Able to set and achieve goals, and work well under pressure.

RELEVANT EXPERIENCE

1990–present CIRCUIT RIDER PRODUCTIONS, Walnut, CA (Nonprofit, integrating environmental management and vocational training)

Division Manager, Vocational Training, 2 yrs.

- Managed implementation of the Vocational Training Division portion of long-range plan:
  - Developed time line and short-term objectives;
  - Defined and delegated tasks;
  - Revised long-term plan as needed.
- Served on agency strategic planning committee, developing 5-year long-range plan, addressing issues of:
  - Current and projected financial and physical resources;
  - Current and proposed educational and environmental services.
- Developed and oversaw $350,000 budget for vocational training division.
  - Consistently completed programs on or under budget while meeting program standards.
  - Secured financial support for training programs from local Private Industry Council and other funding sources, through successful proposals submitted annually.
- Lobbied for job training programs, assisting legislators in developing first drafts of youth training bills and keeping them informed of CRP's programs and developments.
- Initiated contact with potential funding sources (local, county and state governments, and community agencies) to promote training programs of CRP.

Project Coordinator, 4 yrs.

- Developed and managed several pilot programs to train youths and adults, including community Conservation Corp program, and VCR repair vocational program.
- Hired, supervised and trained four year-round staff personnel and 12 seasonal assistants.
- Effectively managed program staff by encouraging pride in performance, supporting individual career development, and training staff in problem solving and teamwork.
- Secured national recognition from National Arbor Foundation, for CRP's "Tree Project," named an "exemplary education project" in 1991.

– Continued –

MARLA MOSES
Page two

Marla breaks down her various roles at Circuit Rider. She also illustrates a good general principle at 2 places in her Work History: Briefly explain the nature of the business whenever it isn't obvious.

ADDITIONAL WORK EXPERIENCE

1989–90 **Horticulturist** SONOMA STATE, BOTANIC GARDEN, Cotati, CA
1987–88 **Naturalist/Educator** NATURE CONSERVANCY, Penngrove, CA

EDUCATION, CREDENTIALS & TRAINING

B.A., Environmental Studies & Planning – Sonoma State University, 1990
Junior College Teaching Credential
California Private Post-secondary Administrator & Instructor Certification

Management Training Courses:

– Making Meetings Work	– Team Building
– Conflict Resolution	– Fiscal Management
– Visionary Leadership	– Stress Management
– Time Management	– Personnel Process

MICHAEL HASSID
5029 Grizzly Peak
Berkeley, CA 94708
(510) 389-6552

Objective: Position as Chief Rehabilitation Officer with City Housing Authority.

PROFILE

- Over 5 years of rehabilitation related experience at owner / manager level.
- Committed to the challenge and responsibility of providing quality and cost-effective construction for public housing.
- Excellent professional reputation among city building inspectors, architects, general / subcontractors, owner / builders.
- Extensive experience with new construction, remodeling and rehabilitation, including estimating and subcontractor scheduling requirements.
- Competence with computer cost estimating and job control programs.

PROFESSIONAL CONSTRUCTION EXPERIENCE

Estimating and Bidding

- Accurately and profitably estimated residential rehabilitation projects from foundation to finish.
- Successfully estimated as a subcontractor on plumbing, heating and solar systems throughout the Bay Area.

Contract Planning and Scheduling

- Developed contract language documenting agreed-upon prices, schedules, procedures and responsibilities.
- Reviewed plans and developed schedules with architects, contractors and subcontractors, assuring proper coordination, efficiency and profitability.

Coordinating and Subcontracting

- Successfully coordinated remodeling and rehabilitation projects, each involving 7 or more specialty subcontractors simultaneously, including:
  ...plumbing ...heating ...electrical ...sheet rocking ...roofing.
- Designed and installed plumbing, heating and solar systems in new construction and rehabilitation projects, both residential and commercial.

Job Control and Building Inspections

- Assured subcontractors' strict adherence to projected scheduling and costs, using computer job control programs to monitor construction contract compliance.
- Achieved consistent high quality control in compliance with building codes and inspection standards.

Final Project Wrap-up

- Completed detailed pickup work on all projects, achieving extremely high level of client satisfaction, consistently generating new referrals and repeat business.
- Completed all paperwork, including Notices of Completion, Lien Releases and closing financial statements.

– Continued –

MICHAEL HASSID
Page two

EMPLOYMENT HISTORY

1995–present	**President/Managing Officer**	HASSID CONSTRUCTION, INC., Berkeley
1993–95	**Owner/Manager**	HASSID CONSTRUCTION, Plumbing / Solar, Berkeley
1991–92	**Co-Owner/Manager**	HASSID & BROWN, Plumbing & Solar, Berkeley
1989–91	**Journeyman Plumber**	HASSID PLUMBING, Berkeley
1985–89	**Apprentice Plumber**	Several contractors in Colorado

SPECIALIZED EDUCATION and TRAINING

1993–94 Architectural and Mechanical Drafting classes;
UC Berkeley Extension and City College, SF

1989–90 Solar System & Hydronic Heating Design / Installation;
Vista College, Berkeley

1985–88 Plumbing Apprenticeship Program;
Red Rocks Community College, Denver, CO

BRENDA GILBERT

49006 LaCrosse Road
Dayton, Ohio
(338) 919-7887

Objective: Position as Management Services Officer.

SUMMARY OF QUALIFICATIONS

- Strong background in financial management and control, using the latest computer technology.
- Highly effective in space development, management and control.
- Technical knowledge in contract preparation and negotiation.
- Skilled in job structuring and resource allocation under tight budgetary controls.
- Resourceful in developing contacts and information sources.
- Talent for picking the right people for the job.

PROFESSIONAL EXPERIENCE

1987–present **Management Services Officer**
University Business School, Dayton OH

Financial Management & Control • Systems Development

- Administered $18 million budget with full signature control.
- Developed highly effective internal cost reporting system using front line computer technology.
- Accurately forecast short-term and long-term financial needs involving complex and speculative base of resources.
- Assessed computer application needs in major change to computerization.

Space Development & Allocation

- Effectively redesigned and reallocated limited classroom and office space to meet constantly changing priorities.
- Located highly desirable new locations and negotiated favorable contracts for MBA Programs at locations throughout the state.
- Developed new building project planning guides.
- Collaborated with architects and interior designers on new facilities for:
  –Career Planning and Placement Center –Computer Center
  –tiered classrooms –special events rooms –major conference areas.

Personnel

- Assessed and managed personnel needs for department of 300 professionals and support staff:
  –established organizational chart;
  –determined specific job responsibilities;
  –allocated staff resources;
  –supervised senior administrative staff;
  –arbitrated personnel disputes;
  –hired, evaluated and terminated staff;
  –established personnel guidelines.

– Continued –

1976–86 **Assistant to Director** External / Internal Affairs,
Ohio Continuing Education of the Bar – Dayton, OH

Financial Management & Control

- Oversaw $9 million budget of this self-supporting organization.
- Submitted successful grant applications for funding of:
  - Development of OEO-funded educational program for neighborhood legal offices;
  - Legal / educational project for mental health social workers.

Programming & Special Projects

- Implemented simultaneous professional programs statewide (approximately 200 programs per year):
  –scheduled programs –selected participants –developed program content;
  –drafted brochures, program handouts, participant guidelines, presenter guidelines; –coordinated physical logistics and staffing.
- Acted as liaison between University and State Bar, monitoring operational agreement.

Personnel

- Recruited high caliber personnel: attorneys, editorial staff, computer operators, administrators, clerks, secretaries, non-professional support staff.
- Set up criteria for new profession of paralegals, as management consultant to lawyer groups.

Systems Development

- Streamlined sales and enrollment accounting procedures for rapidly expanding CEB unit.
- Designed effective new packaging and promotion that increased sales of legal how-to books.

EARLIER PROFESSIONAL EXPERIENCE

1973–76 **Legal Assistant** Rogers, Ryan and Luce, attys. – Dayton
1970–73 **Sales Representative** Gorman Navigation Co. – Kansas City

EDUCATION

B.S., Chemistry / Psychology – Kansas State University

Additional training:
Dayton Law School: Contracts, Torts, Property, Tax
University & Business training schools: Management; Computing & Finance; Communication Skills; Psychology.

– References available upon request –

WILLARD TOWNSEND

1990 - 87th Avenue • Oakland, CA 94603 • (510) 213-6567 home

Objective: Position as Executive Director, City Redevelopment Agency.

SUMMARY

- Six years' successful experience in management.
- Professional reputation for prudent fiscal management.
- Thorough familiarity with redevelopment law and process.
- Exceptional skills in assessing business opportunities.
- Involved community leader; inspires confidence and trust.

RELEVANT EXPERIENCE & ACCOMPLISHMENTS

Business Assessment

- Served as business consultant for the Old Oakland Project:
  - Reviewed site and plans for use;
  - Researched other historic rehabilitation projects;
  - Reviewed City of Oakland Redevelopment Plan for overall perspective;
  - Advised general partner on minority equity participation.

Financial Planning

- Developed present and future strategies to increase clients' investment income through procedural and investment methods, including:
  ...municipal bonds ...real estate investment trusts ...oil and gas investments
  ...real estate partnerships (private and public) ...mutual funds.
- Identified an under-utilized tax deferral system, enabling corporations to increase their investment profits.

Project Development

- Assessed ambitious business plans for a new San Francisco night club, to be the third largest in North America:
  - Evaluated the site; investigated the backgrounds of potential partners;
  - Met with accountants and attorneys to review and document financial projections;
  - Researched plans of similar ventures throughout Europe and the eastern US;
  - Made extensive presentations to investors, raising large amount of capital.

Public Relations/Community Leadership

- Won support for a development project, successfully responding to citizen complaints and convincing city agencies and residents of its benefits to the city.
- Successfully mediated disputes and communication breakdowns between businesses and organizations, clarifying issues and recommending solutions.
- Provided leadership for community service projects, including:
  - Rotary Club of Oakland, World of Work Program;
  - Oakland Parents in Action, anti-drug club;
  - Volunteers of America Bay Area, Inc. Board of Directors, Director of Enterprise.

EMPLOYMENT HISTORY

1991–present	**Owner/Principal**	TOWNSEND & ASSOCIATES, Investments, Oakland
1991–93	**Account Executive**	XEROX BUSINESS MACHINES, INC., Oakland
1990–91	**Life Underwriter**	MUTUAL LIFE INSURANCE CO., Sacramento
1987–90	**Admin. Assistant**	CA. DEPT. OF REHABILITATION, Sacramento

EDUCATION & TRAINING

B.A., Political Science - University of California, Davis
College for Financial Planning, Denver, CO

Rita explains, in her employment history, what "Seventy-Seven, Inc." is about.

RITA L. SASAKI

1990 Norwood Street ● San Francisco, CA 94123 ● (415) 672-8111

Objective: Position as Executive Assistant in an international organization focusing on trade and business development in the Pacific Rim countries.

PROFILE

- Sharp business sense with extensive background in both Asia and the U.S.
- Diplomatic and tactful with professionals and non-professionals at all levels.
- Exceptional communication and interpersonal skills; effective negotiator.
- Analytical and versatile; able to maintain a sense of humor under pressure.
- Poised and competent as a professional business representative.
- Readily transcend cultural and language differences.

PROFESSIONAL EXPERIENCE

Knowledge of East/West Business

- As Project Manager for Seventy-Seven, Inc., served as liaison between parties in high-rise development, coordinating with:
  - American architects and construction firms;
  - Chinese entrepreneur in Hong Kong;
  - British bankers and solicitors in Hong Kong;
  - American and Hong Kong brokers;
  - Hawaii-based American mortgage bankers, attorneys and accountants.
- Worked in Tokyo as executive assistant to the manager of Oriental Exporters Ltd., a New York-based import/export firm.
- Organized and led international tours to China and other parts of the Orient.

Coordination/Problem Solving

- Monitored progress of a $26 million construction project in Hawaii to meet contracted completion date. Mediated job-site conflicts among subcontractors.

Writing & Presentation

- Made oral and written presentations to bankers on financial status and work progress, to get approval and funding for Seventy-Seven's development projects.
- Prepared and delivered briefings for sales agents and brokers.
- Presided over weekly project meetings with contractors and architects.
- Designed forms, correspondence, reports; created and maintained filing systems.

EMPLOYMENT HISTORY

1995–present	**Realtor**	REYNOLDS & CO. REALTY – Hawaii
1991–95	**Project Manager**	SEVENTY-SEVEN, INC. real estate development – Hawaii
1986–90	**Tour Manager**	CULTURAL TOURS, INC. – Hawaii, Hong Kong, Canada
1983–85	**Office Manager**	LUM AND LUM ASSOCIATES architectural firm – Hawaii
1980–83	**Financial Sect'y.**	THE HAWAII CORP. holding company – Hawaii
1978–79	**Executive Sect'y.**	ORIENTAL EXPORTERS (JAPAN) LTD. – Tokyo

EDUCATION

B.A., Asian Studies – UNIVERSITY OF HAWAII, 1990, Summa Cum Laude

SANDRA HOENIG

890 Piedmont Place • Oakland, CA 94611 • (510) 211-5744

Objective: Management position with Winchell Retail Development Corporation.

PROFILE

- Successful background in both franchise management and multi-unit operations.
- Extensive experience implementing company policy.
- Highly effective and diplomatic trainer and manager.
- Expertise in production of candy, ice cream and baked goods.
- Very good communication skills; able to put people at ease.

EXPERIENCE

1995–present ***Owner/Manager*** HEAVENLY COOKIES FRANCHISE, Berkeley

Management

- Coordinated start-up of a new retail food franchise, soliciting and reviewing bids, coordinating construction, and securing permits and licenses.
- Implemented procedures for cash handling; oversaw daily, weekly and quarterly reporting.

Personnel

- Selected, hired and scheduled a staff of 8-12 responsible part-time employees for Heavenly Cookies store, successfully running the operation with only limited supervision.
- Designed procedures for staff consistency in:
  ...customer service ...product handling and stocking ...equipment and store maintenance.
- Trained personnel in: sales techniques and opening/closing procedures consistent with franchise operations manual; production of candy, ice cream and bakery specialties.
- Evaluated employee performance; made recommendations for promotions or dismissals.

Production

- Located best suppliers of ice cream ingredients and tested their products.
- Produced ice cream in accordance with Heavenly Cookies standard formulas.
- Developed special ice cream formulas to accommodate regional customer preferences.
- Produced, and managed daily production of:
  – baked goods: ...giant cinnamon rolls ...five muffin varieties;
  – candy: ...truffles ...melt-aways ...turtles ...clusters ...fruits, etc.

1990–94 ***Area Manager*** PENTA/EYECARE USA, San Diego

- Coordinated daily operations of optometry offices located at 5 sites, involving staffing, patient scheduling, patient records review, customer services, instructing staff in correct techniques for phone answering, patient check-in, product sales and paperwork.

1989–90 ***Sales Associate*** TARBELL REAL ESTATE CO., San Diego

1988–89 *Student* San Diego State; and real estate classes

1984–88 ***Computer Operator*** NAVAL DATA PROCESSING CENTER, San Diego

EDUCATION

Dietary studies, University of Wisconsin, Madison
Associate Sales License in Real Estate

This unique format combines the features of chronological and functional. This works well when one of your jobs is the only source of your most applicable experience.

Human Resource Resumes

Veronica Silva	Admin analyst/healthcare	32
Tyra Beach	Training specialist, corporate	33
Carol Weitzell	Personnel analyst	34
Marsha Rifenberg	Administrative, human resources	35
Margaret Dwaite	Recruiter/Dept. of Forestry	36
Joyce Meyers	Program director, human services	37
Linda Jamieson	Administrative analyst, development	38
Mary Eddy	Human resources/administration	40
Bill Unger	Benefits advocate	42
Anne Hayward	Human resources trainer	44
Mary Newburgh	Analyst, organizational development	46
Stephen Honda	Administrative analyst, relocation	48

VERONICA SILVA
3804 College Ave. • San Francisco, CA 94124
(415) 411-8785 • (415) 903-7900 (message)

Objective: Health care administrative analyst or program coordinator with an agency or firm providing health management services.

PROFILE

- Highly competent professional, able to work well independently.
- Skill in accurately assessing needs and developing program criteria.
- Dependable and hard working; get along well with colleagues.
- Very well organized and thorough in researching information.
- Effective trouble shooter; can be counted on to get the job done.

PROFESSIONAL EXPERIENCE

Program Coordination/Planning

Coordinated and oversaw Medical Audits program:

- Organized working committees of each medical specialty, including representatives from units' professional medical staff and ancillary services;
- Facilitated meetings to develop criteria for conducting patient care audits;
- Conducted audit research, analyzing patients' medical records, applying the above criteria;
- Trained and supervised MR technicians to analyze records;
- Produced statistical report of audit findings, identifying problem areas;
- Followed through to assure that corrective action was taken on problems.

Analysis/Evaluation

Conducted federally mandated audits aimed at cost containment, focusing on three primary criteria and making preliminary assessments of:

- Appropriateness of patient admission;
- Quality of medical care rendered;
- Accuracy of hospital billing.

PROFESSIONAL EXPERIENCE

1994–present	**Review Coordinator**	WESTERN MEDICAL REVIEW INC., Concord
1990–94	**Patient Abstractor**	UCLA STUDENT HEALTH SERVICES, L.A.
1989–90	**Medical Records Supv.**	CEDARS-SINAI MEDICAL CENTER, L.A.
1985–89	**Medical Records Tech.**	USC/COUNTY HOSPITAL, L.A

EDUCATION

Health & Safety – CAL STATE UNIVERSITY, Los Angeles 1990
A.S., Medical Records – CITY COLLEGE, San Francisco

TYRA C. BEACH

687 Eucalyptus Avenue • San Francisco, CA 94127 • (415) 999-7790

OBJECTIVE **Position as Training Specialist for corporate training programs.**

SUMMARY

- 15 years progressively responsible achievement in training, educating, and marketing.
- Six years' successful supervisory and management experience.
- Demonstrated gift for motivating people toward higher achievements.
- Enthusiastic, high energy, and creative in program development.

RELEVANT EXPERIENCE

Training & Instructional Design

- Developed and conducted successful program in stress management.
- Coordinated and presented an effective program in substance abuse education.
- Taught classes in: Health, Physical Education, Family Life.
- Introduced and facilitated effective Values Clarification program.
- Trained and coached successful athletic teams.
- Trained professional staff members for CPR-Basic Life Support certification.

Management & Program Coordination

- Initiated highly successful Wellness Program in Connecticut public school system:
  - Completed a needs analysis; researched potential programs;
  - Designed program to respond to expressed needs; conducted first annual Health Fair designed to increase awareness of stress-reduction techniques.
- Conducted citywide aerobics program for teachers, increasing participation.
- Reorganized and revitalized Human Sexuality curriculum, increasing its acceptance and effectiveness.
- Created an innovative physical education dance program, improving the teaching environment and resolving logistical conflicts.

Marketing & Communication

- Conducted successful public relations meetings on the controversial school subject of Human Sexuality, greatly increasing parental support of the program.
- Promoted winning attitudes in athletic teams, as well as winning teams.
- Increased student involvement, as sponsor of 100-member Pep Club.
- Lectured on health and fitness for Fairvale, Connecticut citywide physical education clinic and for county Alcoholism Council.

WORK HISTORY

1992–present	**Instructor:** Health/Physical Ed., **Trainer: Aerobics Program & CPR-Basic Life Support II**	FORD HIGH SCHOOL, Fairvale, CT
1984–93	**Instructor**: Family Life, Health, Physical Ed.	JR. HIGH SCHOOL, Fairvale, CT
1979–84	**Instructor,** Physical Ed.	JR. HIGH SCHOOL, Fairvale, CT; DOSS HIGH SCHOOL, Louisville, KY

EDUCATION

B.S., Psychology, Health, Physical Ed.; Western Kentucky Univ., Bowling Green, KY
40 postgraduate credits in Counseling, Psychology, and Dance

One of two resumes for Carol.
See also page 135.

CAROL WEITZELL

1912 Kains Avenue • Berkeley, CA 94702 • (415) 235-3980

Objective: Position as Personnel Analyst

SUMMARY

- 7 years' administrative and analytical experience.
- Strength in creative problem solving.
- Outstanding ability in personnel interviewing and skill assessment.
- Designed and implemented highly successful employee training programs.
- MSW degree, focusing on administration and planning.

PROFESSIONAL EXPERIENCE

PERSONNEL

Staff Development

- Developed an in-service training program for social work staff which increased their professional expertise and theoretical background:
  - Developed a form for self-assessment by workers;
  - Wrote and managed annual training budgets;
  - Contracted with trainers to provide instruction in specific issues in social work;
  - Met regularly with supervisors to evaluate the program on an on-going basis.
- Compiled and edited a comprehensive resource manual instructing social workers in services available in Boston, how to access them, and procedures for qualifying.

Skill Assessment & Supervision

- Interviewed & assessed skills of applicants for positions as Child Welfare Workers.
- Mentored, supervised and evaluated a social worker, resulting in an improvement in her administrative skills and her effectiveness in a large public agency.
- Designed a project for student interns that enabled them to learn about community resources and produce a resource used by social work staff;
  - Supervised their daily work and evaluated their performances.

ADMINISTRATION & PLANNING

- Produced a community social service needs assessment to be incorporated in the budget narrative of a Boston DSS office, synthesizing data from:

...National census ...Child welfare worker survey
...Boston redevelopment survey ...DSS computerized records.

WORK HISTORY

1995–present	***Case Manager***	AREA OFFICE ON AGING, W. Contra Costa Co.
1995–present	***Social Worker***	CONTRA COSTA COUNTY, CA
1993–95	***Photographer***	Self-employed
1990–93	***Program Development***	MASSACHUSETTS DEPT. Of SOCIAL SERVICES
1989–90	***Administrative Analyst***	"
1988–89	***Child Welfare Worker***	"
1987–88	***Student Intern***	"

EDUCATION

MSW, Boston University – SRS fellowship in Child Welfare, 1988

See Marsha's cover letter on page 269.

MARSHA RIFENBERG
12 Sherwood Avenue
Oakland, CA 94611
(510) 797-2131

Objective: Administrative position in Personnel and Human Resources.

HUMAN RESOURCES

EMPLOYMENT HISTORY

1994–present **Human Resources Rep.** JENNER ENGINEERING – Oakland

Human Resources Administration
- Recruited, screened and interviewed applicants for exempt positions.
- Designed and implemented a multi-purpose orientation for newly hired employees.
- Interpreted Human Resources policies and procedures to management.
- Resolved employee grievances, avoiding potential lawsuits against the company.
- Coordinated special projects such as the United Way Campaign, Pre-Retirement Seminars and Campus Recruiting Program.

Training & Development
- Conducted corporate training needs assessments.
- Facilitated Quality Circle/Task Force meetings; trained QC leaders and facilitators.
- Provided employee awareness seminars on topics such as alcohol and drugs.
- Planned, delivered and evaluated management/employee development programs.

Employee Assistance Program Coordination
- Assessed employee problems and made referrals to appropriate treatment.
- Coordinated and developed internal communications and promotion of the EAP.
- Advised managers and supervisors on the handling of troubled employees.

1992–93 **Teacher** PERALTA ELEMENTARY SCHOOL, Oakland

1990–present **Mediator** PRIVATE PRACTICE, Oakland
- Conducted effective mediations in the Bay Area over the past 7 years.
- Mediated and arbitrated for San Francisco's Community Disputes Services.
- Successfully headed off a court battle between disputing business partners, negotiating an amicable dissolution of the business and an acceptable financial settlement.

1986–1989 **Office Manager/Instructor** CENTER FOR HUMAN GROWTH, Berkeley
- Researched and evaluated community resources to ensure quality referrals.

EDUCATION, TRAINING & CREDENTIALS

M.S., Educational Psychology – California State University, Hayward, 1995
Chemical Dependency Studies Certificate, CSUH, 1995
Arbitration/Mediation Training, American Arbitration Association, 1995
Multiple Subjects Teaching Credential, CSUH, 1990
B.A., English – Ohio Wesleyan University, 1978

MARGARET DWAITE
1224 Seymour Avenue
Elmira, CA
(123) 456-7890

Objective: Recruitment position with Department of Forestry

SUMMARY

- Extensive **successful experience in recruiting** with many ethnic groups.
- 12 year background in **law enforcement, specializing in investigation.**
- Dedicated, hard-working, and responsible professional; awarded Certificate of Appreciation for **exceptional performance** by Inspector General's office, USDA.
- Quick-thinking, intelligent, able to make on-the-spot independent decisions.

RELEVANT EXPERIENCE

1994–present Employment Specialist — CITY OF ELMIRA, CA
- Assess applicants for potential employment, matching them to employers' criteria for interpersonal skills, technical expertise, and learning potential.
- Achieve an excellent record for successful matches in assisting employers to meet affirmative action goals.
- Conduct community outreach, developing a broad knowledge of area resources:
  – Spoke to community organizations to introduce the "Hire Elmira" program.
  – Recruited individuals at local job fairs.

1992–93 Project Administrator — PORT OF ELMIRA, CA
- Developed and oversaw a highly effective training program, providing a unique opportunity for disadvantaged youth to develop marketable skills and abilities.

1988–91 Investigator — U.S. DEPT. OF AGRICULTURE
- Conducted hundreds of undercover fraud investigations for food stamp program.
- Testified effectively in court, winning judgements against abusers.
- Recruited, hired and trained over 70 investigative aides, of diverse ethnic backgrounds, to assist in undercover investigations.
- Collaborated with other agencies, providing technical advice and investigative support.

1984–87 Full-time student
- Worked as part-time Investigative Aide for the Department of Agriculture.

1978–83 Eligibility Technician — ELMIRA CO. SOCIAL SERVICE DEPT.
- Screened applicants on complex eligibility criteria for a range of social services.

1982–present Administrative Officer — U.S. COAST GUARD RESERVE
- Supervise all administrative functions: payroll, records maintenance, etc.
- Oversee training programs and directly trained over 800 reservists.

1975–78 Deputy Sheriff — ELMIRA COUNTY
- Interviewed and classified inmates to determine appropriate job assignments.
- Prepared comprehensive reports on investigations and crimes.

EDUCATION & TRAINING

112 college units in •Social Science •Business •Criminal Justice
Management and Leadership Training, U.S. Coast Guard
Peace Officers' Standard & Training Certificate (POST)

JOYCE MEYERS

2400 California Street, #203
San Francisco, CA 94109
(415) 702-3939

OBJECTIVE

Position as program director/coordinator/advisor in a human services setting.

SUMMARY OF QUALIFICATIONS

- Demonstrated talent for directing and supervising staff, achieving balance between task needs and employee needs.
- Highly effective in analyzing work flow and communication patterns, to maximize effectiveness of the work team.
- Skilled and confident in organizing start-up phase of new projects.
- Degree in human relations; 12 years' professional experience.

PROFESSIONAL EXPERIENCE

Staff Development & Supervision

- Directed staff of 4–30, including counseling, support, security, and clerical personnel in several work environments: school, counseling center, community treatment centers, and business.
- Trained, supervised and evaluated staff, enabling them to improve skills and achieve work objectives.

Program Coordination

- Designed a highly successful and innovative parent-child relations program:
  – Assessed the needs of the parents, and the staffing housing resources available;
  – Analyzed the program design to anticipate and minimize problem areas;
  – Developed guidelines for participation of parents and staff.
  – Evaluated results of a trial run before full implementation.
- Coordinated the complex logistics of opening two residential treatment centers, including hiring and training staff, and supervising set-up of physical facilities.

Advising/Counseling

- Mediated to identify and resolve conflicts between staff members, clarifying work relationships and alleviating communication problems.
- Counseled individuals to identify and achieve personal objectives.
- Counseled youths and adults in group on problem solving and crisis resolution.

WORK HISTORY

1995–present	**Coord. of Volunteers**	GOLDEN GATE NAT'L. PARK ASSOC., SF
1986–94	**Superintendent**	COMMUNITY TREATMENT CTRS., Oklahoma City
1984–85	**Counselor**	GROUP PROCESSES, counseling center, Oklahoma City
1983	**Project Director**	URBAN LEAGUE – Oklahoma City
1982	**Aide to General Mgr.**	SAN FRANCISCO OPERA

EDUCATION

M.A., Human Relations – University of Oklahoma

HUMAN RESOURCES

LINDA JAMIESON
1200 Michigan Drive
Cortland, NV
(415) 616-5003

OBJECTIVE

Position as Administrative Analyst in a Development Office.

SUMMARY OF QUALIFICATIONS

- Experience in planning and supervising fund-raising events.
- Ten years' demonstrated ability to recruit and motivate volunteer fund-raisers.
- Seven years' experience planning and organizing official functions for the President of the University of Nevada.
- Public Relations experience in a variety of settings.

PROFESSIONAL EXPERIENCE

FUND-RAISING

- For six years, organized volunteers and assisted in planning fund-raising events for the University Art Museum.
- Assisted in the identification of corporate and family foundation prospects for a $500,000 capital campaign for Cooperative Extension.
- Reviewed direct-mail solicitation copy.
- Represented the Division of Agriculture & Natural Resources, filling in for supervisor at UN Development Officers' meeting; participated in University-wide Development Study requested by the Office of the President.

PUBLIC RELATIONS

- Participated in development of an external relations and governmental relations program for the Division of Agriculture and Natural Resources.
- Planned and organized official functions for the President of the University of Nevada at the Richards Estate and on the campus.
  Functions honored donors, Regents, awardees and executive officers on special occasions, including Charter Day.
- As Information Specialist, educated the general public on specific park regulations, policy regarding wilderness use, and environmental quality.
- Spearheaded planning of conferences and educational tours for Cooperative Extension program.

PROJECT MANAGEMENT

- Chaired organizing committee for a major retirement event for University VP.
- As Budget Assistant, interpreted policy and monitored $400,000 discretionary fund for compliance with fund guidelines.
- Revitalized a Sierra Club outings program, doubling participation by recruiting, training and motivating leaders, and initiating an effective advertising program.
- Helped define Wilderness Use Regulations at Yosemite National Park:
  - Integrated data from an inventory of resources and an environmental quality study;
  - Developed and implemented the park's wilderness permit system.

– Continued –

PROFESSIONAL EXPERIENCE, continued

WRITTEN & VERBAL COMMUNICATION

- Conceptualized and wrote descriptive brochure on the Division of Agriculture and Natural Resources.
- Developed extensive, quality network of people resources for Division of Agriculture and Natural Resources.
- Organized and conducted numerous meetings and retreats of association boards and clubs.
- Delivered program status reports before 50-member national Sierra Club committee, successfully persuading officials to continue support of a threatened program.

EMPLOYMENT HISTORY

1996–present	**Space Planning Assistant**	Budget & Space Planning, University of Nevada
1995–96	**Special Project Analyst**	External Relations, Div. of Agriculture & Natural Resources, Offc. of the President, Univ.of Nevada
1992–95	**Budget Assistant**	Budget & Space Planning, University of Nevada
1990–present	**Co-owner/Trip Leader**	Destinations Inc., White-Water Rafting Company
1987–94	**Administrative Assistant**	President's Office, University of Nevada
1986–87	**Computer Coder/Editor**	Reilly Marketing Research Corp., SF
1984–85	**Information Specialist**	US Natl. Parks Service, Yosemite National Park

CURRENT COMMUNITY LEADERSHIP / PROFESSIONAL AFFILIATIONS

1995–96	Elected To Board, University Art Museum Council
1993–96	Elected Sub-Committee Chair, Sierra Club National Outings
1986–96	Leader For Sierra Club National Outings
1994–95	Elected President, Cortland Horsemen's Association
1996	Special Assistant To President For Range Management, Cortland Horsemen's Association

EDUCATION

B.S., (Honors) Conservation of Natural Resources – University of Nevada, 1984

Management Training

"Mid-Management Mentorship;" "Functional Prose for Analysts and Administrators;" "Preparing For Successful Fund-raising;" "Fund-raising Techniques;" Accounting

MARY EDDY

121 Paris Lane • San Mateo, CA 94402 • (415) 771-6163

OBJECTIVE

Position in administration or human resources,
emphasizing analysis, planning, writing, problem solving.

PROFILE

- Outstanding skills in analysis, strategy and planning.
- Proven ability to create and manage a results-oriented team.
- Successful liaison between departments and companies.
- Able to meet demanding time goals.
- Experience in clear and effective business communication.

ADMINISTRATIVE / HR EXPERIENCE

1994–present — ***Human Resources Associate*** — BIOTECH, San Francisco

- Streamlined procedures to generate cash management document:
  - Rewrote job descriptions for greater efficiency with fewer personnel;
  - Retrained clerks in gathering raw data;
  - Designed and implemented a system of quality control.
- Supervised and coordinated major project of assembling benefits information packets, and distributing to 1000 Biotech employees:
  - Assisted manager in determining content of packets;
  - Oversaw two clerks assembling materials;
  - Arranged for printing, photocopying and distribution.

1990–93 — ***Senior Supervisor*** — SOUTHERN PACIFIC TRANSPORTATION, SF

- Developed strong liaison network with staff in various SP department's subsidiaries and banks, to effectively resolve financial/operational problems.

1989 — ***Crew Leader*** — CENSUS BUREAU, San Mateo

- Managed team of 24 federal census takers handling over 3000 forms:
  - Interviewed, hired, trained and counseled all census takers;
  - Broke down and assigned specific areas;
  - Reviewed and approved completed census forms.

1984–88 — *Student* — CORINTHIAN COLLEGE, San Mateo

1984–88 — *PT Salesclerk* — I. MAGNIN, San Francisco

1982–83 — ***Administrative Assistant*** — NOMURA SECURITIES INTERNATIONAL, SF

- Bridged cultural and language differences successfully:
  - Integrated American and Japanese business office procedures in U.S.-based office of Japanese financial securities firm;
  - Mediated a communication breakdown between American and Latin American military representatives.

– Continued –

ADMINISTRATIVE / HR EXPERIENCE

1979–81 — ***Salary Allotment Clerk*** — EQUITABLE LIFE, San Francisco

1976–78 — ***Educational Services Officer*** — US NAVY, Pensacola, FL

- Re-evaluated work assignments and job descriptions for staff of 5 to distribute tasks more efficiently and equitably.

1978–94 — ***Department Manager*** — US NAVAL RESERVE, Personnel & Admin. Units

- Researched, wrote, and produced a 100-page Procedures Manual covering personnel, training, and administrative issues.

EDUCATION

B.A., Sociology – CREIGHTON UNIVERSITY, Omaha, NE, 1976
Certificate in French – UNIVERSITY OF GRENOBLE, France, 1982

Bill Unger

3056 Hillegass Avenue
Berkeley CA 94705
(510) 845-7418

OBJECTIVE

A position as Benefits Advocate with Homeless Action Center

SUMMARY

- Two years experience as a volunteer Benefits Advocate with HAC.
- Ready to make the transition from volunteer to full-time staff position, carrying a full caseload.
- 100% commitment to doing whatever it takes to secure benefits for clients.
- Strong belief in every human's right to a basic level of dignity and security in their daily life.
- Degree in Psychology, with minor in Sociology.
- Stable lifestyle in Berkeley; here "for the long haul."

RELEVANT EXPERIENCE

CASE MANAGEMENT

- Strive to **maintain regular contact** with each client in order to deal with problems as they come up and remind them of upcoming appointments such as consultative exams, drug and alcohol treatment commitments, and court appearances.
- Maintain a thorough, **up-to-date computerized log** on each client.
- **Promptly respond to all client-related communications** from outside agencies.
- In the past 20 months, gained proficiency in WordPerfect enabling me **to handle my own caseload correspondence** without help.

REFERRALS — COMMUNITY SERVICES

- Developed familiarity and **rapport with allied social service agencies.**
  –Arranged for an immediate therapy appointment for a client who arrived in a depressed, suicidal state, and **personally accompanied him to Berkeley Mental Health. Collaborated with the therapist** in deciding on a 72-hour observation period at John George Pavilion.
  – Developing my own **"important contacts" file** of people in agencies with whom I have had success in the past.

KNOWLEDGE OF BENEFITS PROGRAMS

- Determined eligibility of clients for ...General Assistance
  ...Food Stamps ...SSI
- Developed an **understanding of "how the system works"** and some of the ways to effectively work within that system.

– Continued on page two –

Bill Unger

Page two

RELEVANT EXPERIENCE (cont.)

SENSITIVITY & COMMITMENT TO CLIENTS

- **Successfully secured SSI benefits** for my last two clients at Reconsideration. The most recent case required last-minute extensive documentation of longitudinal evidence. I produced a 6-page letter, working til midnight just before leaving on vacation, since I realized **this case would be lost without my timely intervention.**
- Developed a **sensitivity and ease** in working with **ethnically diverse clients** at HAC.
- **Personally accompanied my client** on public transportation to SSI consultative exam, to offer needed support and ensure his appearance at CE.

INTERVIEWING

- Developed **sensitivity in my interviewing techniques**, for example...
  –Take the time to explain the process to clients and my role in it;
  – Let them know I am there for them;
  – Tune in to their responses and "back off" when appropriate.
- Effective in **gaining trust and drawing out relevant personal information** that is crucial to preparing disability reports, drug and alcohol questionnaires, etc. that directly relate to securing benefits.

RELEVANT WORK HISTORY

1992–present	*Benefits Advocate*	**HOMELESS ACTION CENTER** (volunteer/part-time)
1986–present	*Taxi Driver*	**DE SOTO CAB CO, San Francisco**
1989	*Food Server*	**ST. ANTHONY'S DINING HALL,** (volunteer)
1985	*Taxi Driver*	**YELLOW CAB, San Francisco**

EDUCATION

B.A., Psychology (minor in Sociology) - University of Arkansas at Little Rock. 1980

HUMAN RESOURCES

ANNE HAYWARD
1224 Colusa Avenue
Berkeley, CA 94707
(510) 527-6069

Objective: Position as Human Resources trainer: management & supervision.

QUALIFIED BY

Eight years' experience in adult training; expertise in adult learning theory:
- Designed courses and evaluation techniques.
- Monitored and trained trainers.
- Delivered over 300 hours of adult training.
- Recruited presenters and trainers for 100 programs.

RELATED ACCOMPLISHMENTS

- Designed and presented hour-long weekly orientation program for career development organization; doubled membership.
- Designed five courses and all materials to train over 500 adults in job search techniques.
- Monitored and evaluated 30 trainers for more than 100 programs; instituted quality control methods.
- Developed and delivered public seminars for libraries, police departments, and women's organizations.
- Completed 50-hour advanced course in interpersonal communication and conflict negotiation.

PROFESSIONAL EXPERIENCE

1994–present **Program Director**, Career Options, career development organization, San Francisco, CA
- Designed quarterly programs of 15-20 workshops and panels.
- Wrote course descriptions for quarterly calendar of events.
- Trained trainers and monitored classes.
- Supervised volunteer and counseling staff.
- Managed public relations program.
- Presented workshops.
- Recruited employers and professionals to participate in monthly programs.

1983–93 **Community Organizer**, Berkeley, CA
Concurrent with graduate studies at CSU/Hayward:
- President of three parent/teacher organizations;
- Appointed member of Citizens Task Force to elect a new School Superintendent;
- Researcher/Writer/Editor, Crime Prevention Newsletter;
- Election campaign steering committee/school board supervisors;
- Elected member, BUSD Teachers Corp. Program;
- Earned secondary teaching credential, U.C. Berkeley.

– Continued –

Chronological/
accomplishments
combination.

ANNE HAYWARD
Page two

PROFESSIONAL EXPERIENCE, continued

1976–83 **Head Librarian & Children's Librarian,** Wayne County Library System, MI; San Francisco Public Library; San Francisco, CA.

EDUCATION & CREDENTIALS

M.L.S., 1976, University of Michigan
B.A., 1975, Michigan State University, Honors College
Secondary Teaching Credential, 1992, University of California, Berkeley
Bay Area Writers' Project, University of California, Berkeley

AWARDS

Teacher of the Year, 1992
– Graduate School of Education, University of California, Berkeley

ADDITIONAL COURSE WORK

- Training and Human Resource Development, 1993, UC Berkeley Extension
- Interpersonal Communication, 1992, UC Berkeley
- How to Make Meetings Work, 1995, Interaction Associates, Community Training & Development
- Conflict Negotiation, 1995, Community Training & Development
- Assertive Communication, 1995, Community Training & Development

MARY NEWBURGH
2345 Piedmont Road
El Sobrante, CA 94803
(510) 666-3267

Objective: Analyst position, targeting productivity or organizational development.

PROFILE

- Strong analytical skills, with exceptional attention to detail.
- Successful in establishing productive work relationships.
- Thoroughly explore all avenues and options in solving problems.
- Positive, professional attitude; committed to excellence.
- Outstanding presentation skills, both written and spoken.

EMPLOYMENT HISTORY

1993–present **Management Analyst** U.S. Army Headquarters, San Francisco

Productivity/Planning

- Recommended procedural changes (80% of which were adopted) based on a work productivity study, and developed highly effective presentation outlining the results.
  - Served consistently as study team's preferred presenter, delivering all team briefings to top management, using audio visual aids.
  - Delivered informational talks on study results and newly adopted procedures, to:
    ...employee groups ...administrative units ...office personnel ...managers.

Problem Solving / Needs Analysis

- Developed wide ranging recommendations for cost-effective solutions:
  - Decreased shop labor costs 21% through personnel reallocation;
  - Improved inventory control through establishing optimum quantities for stock on hand, and a regular schedule for inventory;
  - Improved personnel accountability by redefining responsibility for essential functions.
- Streamlined motor pool's administrative procedures, saving money, space and time by identifying and eliminating needless steps, and computerizing the process.
- Created first-ever comprehensive picture of base road facilities, quantifying and consolidating historical data, maps, blueprints and past studies.

Cost/Benefit Analysis & Organizational Development

- Saved over a half-million dollars for government offices:
  - Gathered and analyzed cost data and documentation provided by suppliers;
  - Made formal presentation to upper management;
  - Recommended conversion from lease to purchase of word processing equipment.
- Developed working knowledge of evaluation techniques:
  - work sampling; – operational audit; – group timing technique;
  - engineered and non-engineered standards; – technical estimate.

– Continued –

Mary shifts the emphasis by detailing her on-the-job management training courses in the Army, while down-playing the French and Literature background.

MARY NEWBURGH
Page two

EMPLOYMENT HISTORY, Continued

1992–93 **Administrative Assistant** U.S. Army Headquarters, San Francisco

1989–91 **Interpreter/Tour Guide** Freelance/self-employed – San Francisco

- Interpreted negotiations and translated documents regarding the marketing of an imported French soft drink.
- Provided translation for technically complex quality control meetings.

1987–89 **Tour Director/bilingual** HOLIDAY TOURS – San Francisco

1985–87 **Sales Associate** LACHMAN & EMERIC, Stockbrokers – S.F.

EDUCATION & TRAINING

M.A., French – SAN DIEGO STATE UNIVERSITY, 1991

B.A., Comparative Literature – SAN DIEGO STATE UNIVERSITY, 1983

Management Engineering courses:

- Planning & Conducting Management Audits
- Administrative Systems Analysis & Design
- Economic Analysis for Decision Making
- Work Methods and Standards
- Organization Planning
- Work Planning & Control Systems
- Human Behavior in Organizations
- Efficiency Reviews

Stephen uses the functional format to highlight and maximize his minimal experience in the new target job. See Stephen's other resume on page 245.

STEPHEN R. HONDA

788 Manada Avenue • Oakland, CA 94612 • (510) 890-6443

Objective: Position as Administrative Analyst, Relocation Division,
Community Development Department, City Of San Pablo.

SUMMARY

- Conducted extensive research and analysis of RE prices and availability.
- Designed and implemented surveys on relocation needs.
- Adept in counseling residents on relocation options and responsibilities.
- Strength in innovative program planning, funding, and analysis.

COMMUNITY DEVELOPMENT EXPERIENCE

Research and Analysis

- Conducted survey and established vacancy rate in high-density Oakland neighborhood, and presented findings to City Council and Planning Commission.
- Developed and administered a survey of business and relocation needs in redevelopment area.
- Researched and compared Concord condominium prices.
- Extensively studied and documented mortgage availability and distribution in City of Oakland via: ...personal interviews ...documents collection ...correspondence.
- Researched laws governing transfer and write-down of land from public to private ownership in New York City, and submitted detailed report advising City Attorney of contract requirements.

Relocation Counseling

- Counseled residents on timetable, options and responsibilities in conversion and eviction procedures.
- Wrote description of eligibility requirements for Concord home ownership assistance program and explained requirements and limitations to potential home buyers.

Program Administration

- Set budget priorities for community development projects, incorporating community input.
- Completed annual CDBG performance report.
- Surveyed and compared methods of saving and leveraging city's CDBG funds.
- Applied to foundations for historic preservation funds.

WORK HISTORY

1994–present	***Classroom teacher***	Oakland and Richmond public schools
1993	***Office Manager***	Pacific Car Rental, Oakland
1991(summer)	***Administrative Asst.***	**Community Development Dept.**, Concord
1990–1992	Full-time grad student	Columbia University, NYC
1990	***Researcher/Writer***	**Task Force on City-owned Property**, NYC
"	***Library Clerk***	UC Berkeley
1989	***Housing Advocate***	**Oakland Citizens Committee for Urban Renewal**
1988 (summer)	***Housing Intern***	**Calif. Dept. of Housing & Community Development**

EDUCATION & TRAINING

Completed graduate course work in Urban Planning, COLUMBIA UNIVERSITY, NYC
B.A., Geography – UC BERKELEY

Administration & Coordination Resumes

Christine Gade	Project supervisor/construction	50
Edith Levenson	Public information officer	52
Deborah Elstad	Program director	54
Christie Keller	Administration, performing arts	56
Anne Fullbright	Events coordinator	57
Lani Simpson	Admin/manufacturing support	58
Ellen Cummings	Project assistant	59
Heather Arnold	Program management	60
Grace Flanders	Fund raiser, development	62
Shirley Krenz	Project management/customer service	64
Mary Quinlan	Administrative, international trade	65
Nancy Chicago	Assoc director/college admissions	66
Susannah Holt	Volunteer coordinator, humane society	67
Gelia Thornton	Purchasing/inventory control manager	68
Roberta Swan	Program management, social service	69
Rose Ellington	HMO member relations manager	70

CHRISTINE GADE

998 Vincente Avenue • Reno, NV 89511 • (702) 615-6755

OBJECTIVE

Project Supervision and Purchasing for new construction or remodeling of a multi-unit hotel, casino, office or apartment complex.

SUMMARY

- Highly skilled in purchasing, and fine-tuning of FF&E* specifications.
- Proven record for maintaining schedules; never missed a deadline.
- Effective negotiator and decision maker; direct, clear and confident in managing multi-million dollar expenditures.
- Building Department background; broad knowledge of building codes and fire codes.
- Dedicated professional attitude, committed to getting the job done.

SKILLS AND EXPERIENCE

Coordination/Supervision

- Successfully coordinated elements of interior finish work, maintaining project schedules with minimal impact on hotel operations and revenues.
- Coordinated complex interfacing of subcontractors for complete guest room finishing: carpet layers, painters, electricians, furniture movers, drapery installers, etc.
- Served as emergency project trouble-shooter, interpreting blueprints and making decisions on last-minute design details and changes.

Purchasing

- Purchased guest room, office and public area furnishings:
  - Furniture for guest tower, lobby, casino and convention center;
  - Carpeting, wall coverings, lighting fixtures, drapes, bedspreads.
- Researched and found excellent cost-effective alternatives to specified draperies and bedspreads for the new Benson's Casino 350-room guest tower:
  - Identified inherently flameproof fabric, not requiring annual FR treatment;
  - Met aesthetic and maintenance criteria, greatly reducing cost of upkeep.
- Located best vendor for wall covering materials, reducing the complexity of purchasing and increasing the value-per-dollar, with no sacrifice to aesthetics.

Contract Negotiation/Compliance

- Authored and rewrote specifications, as needed, for bidding purposes.
- Protected owner's financial interests during the bidding process:
  - Maintained the integrity of bidding procedures, with equitable access to information;
  - Warned prospective bidders of the consequences of unethical practices.

– Continued –

CHRISTINE GADE

Page two

SKILLS AND EXPERIENCE *(continued)*

Contract Negotiation/Compliance, continued
- Enforced contract compliance for installation of fixtures, furnishings and equipment, carefully monitoring quantity and quality of both labor and materials.

Knowledge of Codes
- Reviewed specifications in detail, identifying items potentially not in compliance with codes.
- Submitted alternative finishing materials for approval of State Fire Marshal.
- Negotiated with State of Nevada officials to clarify code standards in absence of established guidelines.

EMPLOYMENT HISTORY

1994–present	***FF&E* Administrator***	BENSON'S RESORT HOTEL / CASINO, Reno, NV
1993	***Project Assistant***	BENSON'S RESORT HOTEL / CASINO, Reno, NV
1989–92	***Administrative Asst.***	BUILDING DEPT., DOUGLAS CO., Minden, NV
1983–88	***Office Manager***	BILOTTI TRUCKING CO., Stockton, CA

(*Fixtures, Furnishing & Equipment)

EDUCATION B.A., English, UNIVERSITY OF OREGON; Eugene, OR

ADMINISTRATION

EDITH T. LEVENSON
400 Dolphin Drive
Santa Rosa, CA
(707) 215-0033

Objective: Position as Public Information Officer with City of Santa Rosa.

SUMMARY

- Familiar with the City of Santa Rosa; active resident for 23 years; take personal pride in representing my community.
- Impeccable reputation within the community for credibility, professionalism, and dependability.
- Energetic and highly effective leader and team member.
- Skill in public speaking and written communication.

EXPERIENCE & ACCOMPLISHMENTS

SPECIAL EVENTS COORDINATION

- Spearheaded highly successful Annual Community Festival for Santa Rosa, increasing attendance by 30% and income by over 20%:
  - Planned, scheduled and coordinated complex details of the event including entertainment, decorations, publicity, public safety, parking and utilities;
  - Recruited and coordinated nearly 300 volunteers;
  - Obtained all permits and licenses;
  - Coordinated with agencies to avoid schedule conflicts and assure smooth logistics involving transportation, sanitation and easy public access.
- Planned and coordinated numerous annual events, for example:
  - trade fairs;
  - small business award dinners;
  - holiday decoration contest;
  - student recognition dinners;
  - officer installation banquets;
  - general membership meetings.

COMMUNITY PROMOTION

- Spoke before varied audiences on behalf of local social service agencies and officially represented the Santa Rosa Chamber of Commerce.
- Successfully persuaded businesses, agencies and hundreds of individuals to donate time, talent, materials and money in support of community projects.

PUBLIC INFORMATION/ KNOWLEDGE OF THE COMMUNITY

- Advised visitors, potential residents, and employers of the opportunities and characteristics of the community, including history, demographics, cultural offerings, housing, jobs, public service agencies and general community ambiance.
- Actively participated in a wide range of community organizations and programs:
  - Smith Memorial Church
  - Salvation Army
  - Santa Rosa Socialization Center
  - Bay Area Community Service
  - Santa Rosa Unified School District
  - Women's Lunch Network
  - Eden Express Restaurant
  - Santa Rosa Arts Council
  - League of Women Voters

– Continued –

EXPERIENCE & ACCOMPLISHMENTS, Continued

MEDIA RELATIONS

- Successfully solicited national TV coverage for community and project promotion:
  - Persuaded Channel 5 Evening Magazine to choose Santa Rosa over 35 other communities for a Salute to Communities;
  - Convinced Channel 2 that "The Best of Santa Rosa" board game was newsworthy.
- Achieved extensive free coverage in local press, TV and radio for community projects, composing press releases and PSAs, and conducting TV interviews, phone interviews and anchor spots.

EMPLOYMENT HISTORY

1995–present	***Community Promoter and Special Events Coordinator***	SANTA ROSA CHAMBER OF COMMERCE " " "
1995–present	***Promotion Consultant****	UNIVERSAL CUSTOM PROMOTIONS, Cupertino
1993–94	***PR/Fund-raising Rep.***	SANTA ROSA SOCIALIZATION / PRE-VOCATION CENTER
1992–94	***Art Exhibit Coordinator***	EDEN EXPRESS nonprofit restaurant
1986–90	***Adult Ed. Art Teacher***	SANTA ROSA UNIFIED SCHOOL DISTRICT
1976–present	***Art Teacher for children****	Self-employed
1976–81	***Police Matron***	CITY OF SANTA ROSA POLICE DEPT.
1975–77	***Police Radio Dispatcher***	" " "

*concurrent with above

EDUCATION

B.A., Art – UNIVERSITY of NORTHERN IOWA, Cedar Falls, IA
California Teaching Credential, CSU, Hayward
Seminars in Promotion and Marketing

DEBORAH ELSTAD

648 Crescent Street, #133
Oakland, CA 94610

home: (510) 653-9090
work: (510) 645-4000

OBJECTIVE

Position as Program Director / Administrator.

SUMMARY

- Masters in Public Administration; 7 years' professional experience.
- Successful in implementing new programs.
- Exceptional skill in personnel supervision and training.
- Management talent for "seeing the whole picture."
- Effective in budgeting and long-term planning.
- Strength in problem solving and conflict resolution.

PROFESSIONAL EXPERIENCE

1993–present **ASSISTANT DIRECTOR**
OPERATION CONCERN outpatient mental health agency,
PACIFIC PRESBYTERIAN MEDICAL CENTER, S.F.

Program Administration

- Supervised day-to-day operations of the out-patient mental health clinic:
  – Assured that standards were maintained for quality of client service;
  – Constantly observed overall activity in the clinic; redirected staff resources as needed.
- Served as Acting Director in absence of the Executive Director.
- Implemented prompt action to resolve administrative emergencies, such as last-minute budget changes, urgent maintenance or repair of equipment, and staffing shortages.
- Developed agency policy on all issues involving personnel and operations.
- Negotiated and monitored contract with Community Mental Health Services.
- Oversaw and approved selection and purchase of all office equipment and supplies.

Budgeting / Financial Planning

- Assisted Director in developing long-term financial plan.
- Oversaw agency's annual budget of $700,000:
  – Conducted detailed comparison of monthly expense statements and projected revenues;
  – Identified potential cost overruns and collaborated with Director to resolve the problem;
  – Developed annual expense budget, and submitted quarterly financial status statements.
- Conducted complex monitoring of monthly income, in collaboration with agency's accountant, involving revenues from city, state, private foundations, client fees, grants.

– Continued –

DEBORAH ELSTAD
Page Two

PROFESSIONAL EXPERIENCE, Continued

1990–1992 **OFFICE MANAGER /ADMINISTRATIVE DIRECTOR**
OPERATION CONCERN outpatient mental health agency,
PACIFIC PRESBYTERIAN MEDICAL CENTER, S.F.

- Supervised clinical and administrative staff of 23 regarding all administrative activities.
- Interviewed, hired and terminated clerical and administrative staff and maintained their personnel records.
- Successfully mediated conflicts between management and staff, improving internal communications and cooperation.
- Conducted weekly staff meetings and weekly administrative task meetings.
- Negotiated personnel issues, serving as liaison with hospital Personnel Office.

1987–89 **COMMUNITY RELATIONS COORDINATOR**
RIVER QUEEN COMMUNITY CENTER, Guerneville, CA

- Implemented all publicity activities, including preparation of public service announcements, fliers, newspaper articles, and monthly newsletter.
- Spoke before community groups, outlining services provided at the center.
- Organized, publicized and coordinated all community fund-raising events.

ADDITIONAL EXPERIENCE

TEACHING

1988–89 Taught "Interpersonal Criticism," applicable to conflict resolution and communication, at San Francisco State University.

1985–86 Taught "Changing Woman," involving women's changing role in the work world, at Sonoma State University, Rohnert Park, CA.

EDUCATION

M.P.A. (Masters in Public Administration), Golden Gate University; San Francisco, CA, 1983
B.A., Political Science, San Francisco State University, 1977

– References available upon request –

CHRISTIE KELLER

1213 Filbert Street • San Francisco, CA 94123 • (415) 674-5499

OBJECTIVE

Administrative position with a performing/visual arts organization:
Marketing • Development • Program Coordination • Public Relations.

SUMMARY OF QUALIFICATIONS

- Degree in Arts Administration; academic background in the arts.
- Coordinated highly successful PR and fundraising events, applying a natural flair for marketing.
- Experience providing management and direction for nonprofits.
- Special talent for motivating and training volunteer staff.

RELEVANT ACCOMPLISHMENTS

MARKETING & PUBLIC RELATIONS

- Increased public awareness of the economic and political issues involving women's employment, through coordination of highly successful public rally:
  - Prepared local PR releases and packets;
  - Channeled national PR to local media;
  - Represented organization on TV and radio;
  - Served as rally Master of Ceremonies;
  - Developed successful logos and slogans for products sold nationwide.
- Prepared clear and effective press releases for San Francisco Jazz Dance Company.
- Managed media data bank and coordinated mailings for employment service organization.
- Produced videotape for grant application, well received by the California Arts Council (CAC) and later incorporated into SF Jazz Dance Company's marketing.

DEVELOPMENT

- Developed a grant proposal submitted to CAC which was instrumental in placing an arts organization on the priority list for funding consideration:
  - Prepared an appropriate request, researching and consulting with CAC and staff;
  - Wrote comprehensive description of the program and its goals and objectives.

MANAGEMENT & ADMINISTRATION

- Developed an in-depth working knowledge of program development and management principles, through studying 20 resource sharing programs of Bay Area nonprofits;

  Identified the following as among the predictors of successful programs: Participant consensus on goals and objectives; Research of similar programs; Accurate projections of budget and staff needs; Appropriate growth rate.

WORK HISTORY

1994–present	**Board Member/Consultant**	SAN FRANCISCO JAZZ DANCE COMPANY
1993	**Arts Advisory Panel Member**	SAN FRANCISCO FAIR & EXPOSITION
1988–92	**Prog. Coord./Steering Comm.**	WOMEN ORGANIZED FOR EMPLOYMENT
1985–87	**Development Associate**	SAN FRANCISCO BALLET
1985	**Aide to General Manager**	SAN FRANCISCO OPERA

EDUCATION

M.A., Arts Administration – GOLDEN GATE UNIVERSITY, SF
B.A., Cultural Anthropology – UNIVERSITY OF WASHINGTON, Seattle, WA

ANNE FULLBRIGHT
2323 Blake Street
Berkeley, CA 94704
510-540-0707

Objective: Position as Events Coordinator.

SUMMARY

- Successfully coordinated educational events.
- Infectious enthusiasm for promoting programs.
- Excellent at follow-up and detail; extremely dependable.
- Skill in developing cooperative relationships with caterers, managers, etc.
- Tolerant and comfortable with all kinds of people.
- Professional in front of large groups; able to introduce speakers.

EVENT COORDINATING EXPERIENCE

- Set up workshop materials (books, documentation, binders, literature, name tags, pens) for seven computer applications seminars, involving from 60-180 people.
- Signed in registrants, collected/recorded workshop fees, and prepared receipts.
- Opened seminars: introduced speakers, presented schedule, provided general orientation.
- Acted as liaison between hotel staff and attendees.
- Presented Certificates of Completion; collected and processed evaluation forms.
- Updated registration records and distributed accurate Attendee List to each participant.
- Made detailed arrangements for annual student field trips, including transportation, scheduling, soliciting ticket contributions, and adult assistance.
- Organized annual new-teacher luncheons for 200: arranged hotel accommodations, projected costs, selected menu, prepared publicity fliers, greeted and hosted.

EMPLOYMENT HISTORY

1994–present	**Seminar Coordinator**	NEW TECHNOLOGY INSTITUTE, Santa Barbara, CA
1993–94	**Textbook Analyst**	EPIE INSTITUTE Educational Testing, Berkeley, CA
1990–94	**Studio/Business Asst.**	COILLE HOOVEN PORCELAIN, Berkeley, CA
1972–90	**English Teacher**	OAKLAND PUBLIC SCHOOLS, Oakland, CA
1970–72	**English Teacher**	ADAMS COUNTY PUBLIC SCHOOLS, Northglenn, CO

EDUCATION & TRAINING

B.A., English, University of Colorado, Boulder, CO
California public school teaching credential

> If Lani applies for a similar job in another industry, she will change the line in the Summary and the short paragraph about "background in textiles." This way she can create what appears to be a very customized resume for the textile world, with the flexibility to tailor it for other environments.

LANI SIMPSON

5585 Berkeley Hills Way
Berkeley, CA 94706
(510) 555-5732

Objective: Administrative support position with a textile manufacturer, involving •Project Management •Customer Relations •Inter-Staff Liaison.

SUMMARY

- Talent for generating support and cooperation of staff, maximizing and integrating their talents to achieve the priorities of management.
- Able to "think on my feet" and balance many projects at once.
- Proven record of success in handling increasing levels of responsibility.
- Degree in textile design; portfolio of designs available.

– *PROJECT MANAGEMENT* –

- **Set up a special in-house tracking system** for customer repair units.
- **Developed initial plans for regional newsletter** for field service technicians.
- **Organized a recall-and-return program** with product manager for retrofit of customer units, successfully resolving this design problem.
- **Organized all aspects of shipping and receiving** of returned customer goods, **reducing turnaround time 50%** through careful tracking.
- **Filled in for Dept. Manager** during vacations, attending managers meetings, arbitrating disputes, authorizing customer credits, screening potential hires.

– *CUSTOMER RELATIONS* –

- **Reorganized and streamlined customer service filing system** at Brown-Webber for full staff access.
- **Built a reputation as customer service "Answer Lady"** by developing expertise in the product line, responding promptly to clients' questions and requests; take pride in getting problems handled correctly the first time.

– *LIAISON* –

- **Served as management/staff liaison** for **conversion to new software system**:
  – Performed troubleshooting – Trained staff – Recommended improvements.
- **Provided critical management support** for technical managers:
  – Prioritized incoming communications from customers, factory, headquarters;
  – Edited outgoing communications for foreign-speaking technical managers.

– *BACKGROUND IN TEXTILES* –

- **Completed textiles classwork** as required for A.A.S. degree.
- **Independently created fabric designs** for use in apparel.

EMPLOYMENT HISTORY

1992–present BROWN-WEBBER, Optics, San Leandro, CA
Regional Service Administrator/West 1996–present
Supervisor/Central Repair Administrator 1993–96
Group Leader/Central Repair 1993 Apr–Oct
Data Entry/Customer Service 1992–93

1990 summer SPORTS CABLE TV, Oakland, CA – **Customer Service Rep**

1989–90 FUN FABRICS, Albany, CA – **Customer Service Rep**

EDUCATION & TRAINING

B.A., Studio Arts, Montana State University, Bozeman, MT, 1985–89
A.A.S., Fashion Institute of Technology, Carmichael, CA, 1990–91

This flight attendant applies her transferable skills to a new career objective and creates a resume to use while exploring the new field.

ELLEN CUMMINGS

332 Traverse Road • Sausalito, CA 94965 • (415) 392-4455

Objective: Position as project assistant or program coordinator, with a focus on needs analysis, creative problem solving and public relations.

PROFILE

- Strength in anticipating problems and needs before they arise.
- Proven record of excellence and dependability.
- Confident and decisive under stressful conditions.
- Firsthand experience with worldwide range of cultures.
- Talent for getting diverse groups to work well together.

RELEVANT SKILLS & EXPERIENCE

Analysis/Problem Solving

- As Tour Guide, conducted first-hand comparative evaluation of worldwide travel markets and advised individuals and small groups on:
  ...sight-seeing ...reputable merchants and best buys
  ...travel accommodations ...restaurants ...hotel accommodations
  ...skiing costs and skill-level services ...ambiance.
- As Flight Attendant, redesigned food service operations, improving customer satisfaction:
  – Coordinated preparation and delivery into an efficient, smooth-flowing schedule;
  – Upgraded standards of food quality, aesthetic presentation and timeliness.
  – Trained and supervised newly hired flight attendants.
  – Provided individualized services for thousands of airline customers, assessing needs and handling difficult customers effectively.

Organizing/Coordinating

- As student, directed special events for university student social organizations:
  – Chaired membership drive event which increased club membership by 25%;
  – Created and directed promotional skits involving 18-member cast, boosting public support and enthusiasm for sports events and increasing attendance to capacity.

Public Relations

- As Flight Attendant:
  – Developed expertise in working with worldwide range of cultures and languages: European, African, South and Central American, Middle Eastern, Indian, Pacific Basin.
  – Represented United Airlines at public events and as company ski team member.
  – Translated flight announcements into French.

WORK HISTORY

1994–present	***Flight Attendant***	UNITED AIRLINES, San Francisco, CA
	– Won in-flight services Recognition Award twice	
1990–94	***Flight Attendant***	UNITED AIRLINES, New York, NY
1986–90	***Full-time Student***	BUCKNELL UNIVERSITY, Lewisburg, PA
1987	***Tour Guide***	GOLD SEAL VINEYARDS, Hammondsport, NY

EDUCATION

B.A., Art History – Bucknell University, Lewisburg, PA

Additional studies in computer fundamentals and financial accounting

ADMINISTRATION

HEATHER ARNOLD

1200 Circline Drive • Oakland, CA 94610 • (510) 388-1919

OBJECTIVE: Program management position with a corporation or agency, focusing on creative program design, administration and training.

SUMMARY OF QUALIFICATIONS

- 7 years' successful experience in management.
- Creative flair for generating and presenting program ideas.
- Inspire and support others to work at their highest level.
- Ability to prioritize, delegate and motivate.
- Exceptional communication and interpersonal skills.

PROFESSIONAL EXPERIENCE

MANAGEMENT & SUPERVISION

- Planned $250,000 food budget and monitored expenses for Child Care Food Program, maintaining mandatory records and filing for reimbursement.
- Orchestrated complex transportation of 1800 meals per day to 22 sites, from one central kitchen in San Francisco.
- Supervised staffs of up to 90 employees:
  ...interviewing and hiring; ...scheduling;
  ...attendance control; ...evaluating.
- Supervised the training of university student interns.

ADMINISTRATION

- Coordinated all aspects of the nutrition component of Head Start programs for 1000 children in San Francisco and Alameda, involving: ...food service administration;
  ...nutrition education; ...program compliance.

CREATIVE PROGRAM DESIGN

- Revitalized newsletter of Bay Area Dietetic Association, making it the most creative such newsletter in the nation, with a redesigned masthead, new features of popular interest, and reflecting a more professional image.
- Succeeded in winning media coverage for The Body Shop program by writing dynamic proposal for a feature story on childhood obesity.
- Designed innovative promotional campaigns, as intern at New York City PR firm.
- Introduced, for the first time, the use of closed circuit TV as an effective medium for teaching hospitalized diabetic patients how to control their disease.
- Chose topics of high public interest, and located expert speakers, for hospital's community health program promotion.

– Continued –

Heather plans to do some informational interviewing to narrow down her job focus, and then insert a more specific job objective.

HEATHER ARNOLD
Page two

RELEVANT EXPERIENCE, continued

TRAINING & EDUCATION

- Initiated a 5-week nutrition education program for recovering patients in an alcohol and substance abuse treatment facility.
- Monitored parent education program for low income families, assuring appropriate focus on identified nutritional problems of anemia, obesity, high sugar intake, etc.
- Taught weekly classes for children ages 8-12, on behavior modification, self-esteem, parent support, and physical exercise, in a weight control program.
- Trained teachers in effective teaching techniques with pre-school children, and monitored their use of the nutrition curriculum.

EMPLOYMENT HISTORY

1996–present	***Nutrition Consultant***	SAN FRANCISCO HEAD START;
	" "	ALAMEDA HEAD START
	" "	THE BODY SHOP, Oakland, CA
1995	***Clinical Dietitian***	OAK CREEK MEDICAL CENTER, Arlington, TX
1994–95	***Dir. Nutritional Svcs.***	RALEIGH HILLS HOSPITAL, Dallas, TX
1993–94	***Food Supervisor***	ARLINGTON MEMORIAL HOSPITAL, Arlington, TX
1991–93	***Food Supervisor***	SKY CHEFS / AMERICAN AIRLINES, Dallas, TX
1989–90	***Admin. Dietitian***	KNUD-HANSEN MEMORIAL HOSPITAL, St. Thomas, VI

EDUCATION

COLUMBIA UNIVERSITY, New York City; M.S., Education
SYRACUSE UNIVERSITY, Syracuse, NY; B.S., Nutrition & Food Science
American Dietetic Assoc, Reg. #9876543

– References available upon request –

GRACE FLANDERS

390 Forest Blvd.
San Francisco, CA 94115
(415) 214-6771

OBJECTIVE: Position as development officer or fund-raiser.

PROFILE

- Strength in planning and executing successful fund-raising campaigns.
- Understand the philosophy of philanthropy.
- Special talent for organizing and motivating volunteers.
- Committed to and experienced in work that furthers the development of nonprofit organizations.

RELEVANT EXPERIENCE

Annual Giving

- Created first-ever annual giving campaign for Crowell Daycare Center, successfully persuading parents and other supporters to contribute to their maximum capacity.
- Chaired Phone-a-Thon for Bayside Primary School which surpassed projected goal; introduced and implemented several new and successful fund-raising techniques.

Events Planning & Evaluation

- Organized "Book Week" for Bayside Primary School, developing and monitoring the time line, recruiting committee chairs, and greatly improving the record-keeping.
- Chaired committee for Sunrise Day School Auction, a major fund-raiser, projecting staffing and scheduling needs.

Implementation

- Identified and adapted successful fund-raising techniques for a phone-a-thon event:
  – Targeted donor groups and developed specific techniques to solicit each;
  – Recruited and trained volunteers;
  – Accurately tracked pledges and immediately acknowledged gifts.
- Transformed school's "Book Week" from a minor, disorganized event into a successful annual fund-raiser by developing a disciplined, workable organizational structure.
- Identified, trained and scheduled staff of 30 for Sunrise Day School Auction.

Solicitation

- Solicited donations on a one-on-one basis, with a consistently high level of success:
  – Surpassed goals in personally soliciting major contributions from parents for the first-ever capital campaign for Sunrise Day School; raised more money than any other phone solicitor, for three consecutive years, working on the annual giving campaign;
  – Successfully persuaded merchants, service providers and professionals to contribute goods and services to auctions for three private school fund-raisers.

- Continued -

"Family management" is used to fill in the gap in Grace's 'employment history'; in this case we changed it to "work history".

WORK HISTORY

1993–present	***Residence Manager***	Supervising unique home for S.F. Ballet School students
1989–92	Family management	
1986–88	***Job Placement Counselor***	ENTERPRISE FOR HIGH SCHOOL STUDENTS, SF
1985–86	***Coord. of Volunteers***	COMMON CAUSE , San Francisco
1983–84	***Assistant Teacher***	MONTESSORI SCHOOL , LaCrosse, WI
1980–82	Travel in Asia	
1974–79	***Assistant Teacher***	CROWELL DAYCARE CENTER, MI BAYSIDE PRIMARY SCHOOL, NY SUNRISE DAY SCHOOL, CA

EDUCATION

B.A., English Literature, UNIVERSITY OF MICHIGAN
THE FUND-RAISING SCHOOL, San Francisco 1996

ADMINISTRATION

SHIRLEY E. KRENZ

12 Bruenner Square
North Oakfield, CA
(405) 777-1313

Objective: Project management position in Marketing, Public Relations or Customer Service.

SUMMARY

- Enthusiastic, committed, resourceful; can be counted on to get the job done.
- Outstanding record in recruiting, training and motivating employees.
- Successful in negotiating and winning cooperation and support.
- Poised and professional with both top management and support staff.
- Able to pull together and manage all aspects of a complex project.

PROFESSIONAL EXPERIENCE

1990–present **Regional Administration Manager** TAYLOR INSTRUMENTS – Oakfield, CA

Management & Supervision

- Managed implementation of sales office automation:
  - – Met with national management, as local office rep reporting on field requirements;
  - – Hired personnel to operate two computer stations;
  - – Selected and purchased work station furnishings;
  - – Arranged for both in-house training and local tutoring of new staff.
- Hired clerical and temporary sales support staff for regional operations office.

Public Relations/Presentation

- Found HMO speakers to make presentations to employees on the best health plan options.
- Successfully negotiated for hard-to-get facilities, with sales and catering directors of major hotels.

Project Coordination

- Coordinated relocation planning of western regional office, Taylor Instrument Co.:
  - – Researched locations, developing a criteria graph of variables between locations:
    ...costs per square foot ...tenant improvement costs
    ...restaurant/hotel facilities ...proximity to airports and customer base;
  - – Planned physical layout of new offices, working with building architect;
  - – Acted as liaison between company management and real estate representatives.

1985–90 **Office Manager** TAYLOR INSTRUMENTS – Oakfield, CA

1980–85 **Regional Secretary** TAYLOR INSTRUMENTS – Oakfield, CA

1976–79 **Executive Secretary** MANPOWER INC. – Oakfield, CA

EDUCATION

A.A., Supervision – Chabot College, Hayward, CA
Training in: Presentation for Professional Women;
Peak Performance; Assertiveness; Time Management

ADMINISTRATION

Mary aims for an entry level position combining her administrative and travel experience.

MARY QUINLAN
1290 Golden Gate Blvd.
Walnut Creek, CA 94596
(510) 909-6667

Objective: Administrative position in international import trade/shipping.

PROFILE

- Successfully lived, worked and taught in China; thoroughly familiar with Eastern business practices.
- Managed an annual purchasing budget of a half million dollars.
- Experience in negotiating major maintenance contracts.
- Able to oversee large projects, and follow through to completion.
- Exceptionally well organized and highly motivated.

PROFESSIONAL EXPERIENCE

Management & Administration

- Worked in business and communications at Exxon and WKY-TV for 17 years.
- Calculated and wrote detailed financial analysis of subsidiary operation of Exxon stations, assessing the cost-effectiveness of continuing Exxon's car wash operation.
- Coordinated and implemented a massive changeover project, from leaded to unleaded gas delivery:
  - Identified and recruited appropriate maintenance contractors to modify 3000 service stations;
  - Contracted with laboratory testing facilities to provide on-going gasoline analysis;
  - Wrote and managed detailed budget, and negotiated vendor rates.

Business, Trade, International Culture

- Traveled extensively throughout the Pacific Rim, including:
  - Lived in mainland China (one summer);
  - Taught and worked in Taipei, Taiwan.
- Arranged for cost-effective shipping of consumer goods, complying with import regulations.
- Designed and presented classes in Business Communication to department managers at Cyanamid Taiwan.

WORK HISTORY

1992–96	Full-time student	CAL STATE UNIV., HAYWARD; DIABLO VALLEY COLLEGE
1995	***Instructor***, Business Communications	CYANAMID TAIWAN, Taipei
1985–92	***Maintenance Administrator***	EXXON CO., USA; Walnut Creek, CA
" "	***Administrative Assistant***	EXXON CO., USA; Dallas, TX
1974–85	***Secretary***, Advertising Department	WKY-TV; Oklahoma City, OK

EDUCATION & TRAINING

B.A., magna cum laude; major: Chinese Studies; Cal State Univ., Hayward, 1996
A.A., Psychology; Diablo Valley College, Pleasant Hill, CA, 1995

Intensive Chinese language program, Beijing Language Institute; Beijing, China, 1995
Student observer, auditing on-site Asian business development sessions in Taiwan

ADMINISTRATION

NANCY CHICAGO
2326 Trowbridge Road
Lafayette, CA 94549
(510) 777-9090

Objective: Associate Director position in college admissions.

PROFILE

- 17 years' experience in interviewing and evaluating students.
- Self-motivated, confident, high energy manager.
- Effective leader; able to prioritize, delegate and motivate.
- Reputation for taking the initiative and seeing a task through to completion.
- Excellent writing, speaking and counseling skills.
- Committed to higher education and eager to support it.

PROFESSIONAL EXPERIENCE

1995–present **Admissions Counselor** FARNSWORTH COLLEGE, Hawaii (mainland rep)
- Followed up successfully on prospect leads, as admissions counselor for Farnsworth.

1980–96* **Alumni/Admissions Rep** PINE MANOR COLLEGE, Chestnut Hill, MA
*part time, concurrently with positions below
- Recruited applicants to undergraduate program:
  - Visited secondary schools and community colleges, and consulted with counselors;
  - Advised counselors on availability of scholarship money and financial aid alternatives;
  - Interviewed students and maintained on-going communication, in person and by phone;
  - Counseled students on financial aid options and academic requirements.
- Researched, evaluated and presented data before the state legislature in support of special education and ancillary services.

1993–95 **Sales Associate** MACY'S GIFT & CHINA DEPT., Concord
- Earned distinction as Top Salesperson at both Macy's and Hawaii Realtors. Developed substantial and profitable repeat business through extensive follow-up. Exceeded sales goal at Macy's by 160%.

1993 **Fund-raiser/Coordinator** ST. FRANCIS HOSPITAL AUXILIARY (vol).
- Organized and chaired major fund-raising event benefiting St. Francis Hospital, overseeing the functions of all subcommittees.

1982–92 **Co-Owner/Manager** TBS, INC. (4 retail stores), Hawaii
- Co-owned and managed 4 retail stores (gifts and needlework) in Hawaii:
  ...personnel ...training ...sales ...advertising/promotion ...purchasing ...bookkeeping.
- Directly supervised staff of 16-20 sales employees on a daily basis; designed and implemented incentive programs to motivate and achieve sales goals.

1975–82 **Inventory Auditor** Independent contractor for major retailers

EDUCATION

B.S., Education – UNIVERSITY OF IDAHO, Moscow, ID
A.A., Liberal Arts, Pine Manor College

Some people LONG for a job as a train operator. Others, like Susannah, want to get OUT of that field and work with animals! For this big career change, we emphasized only her experience related to the new career goal.

SUSANNAH HOLT
346 Walpert Street
Hayward, CA 94541
(510) 980-6774

Objective: Position as Volunteer Coordinator for Marin Humane Society.

PROFILE

- Over 10 years' of effective public relations experience.
- Demonstrated talent in assessing volunteers' skills and making appropriate placements.
- In-depth experience with pet therapy programs.
- Established and managed a successful pet care business.
- Supervised volunteers at local humane organizations.

PROFESSIONAL EXPERIENCE

Pet Therapy Programs/Volunteer Work

- Implemented a new **pet therapy program** at The Latham Foundation for Human Education, thoroughly researching other programs and selecting the most appropriate features of each.
- Introduced **pet therapy program** to nursing homes:
  - Made initial contacts and described the program benefits;
  - Scheduled visits to nursing homes;
  - Coordinated efforts with Oakland SPCA.
- Served as a volunteer at three area humane organizations.

Communication & Public Relations

- Worked directly with hundreds of **pet care clients**, advising and assessing their pet care needs.
- Effectively handled emergencies and customer inquiries, as BART train operator, earning commendation for **outstanding service to patrons**.
- Mediated between volunteers and staff at the Oakland SPCA to maintain harmonious working relationships and **maximize volunteer job satisfaction**.

Management, Supervision & Training

- Started a **pet care business** from scratch, on my own:
  - Interviewed job applicants, assessed their skills, placed, trained and supervised;
  - Wrote all the contracts, generated billings, followed up on billings.
- Trained new BART train operators.
- **Trained new volunteers** at Latham Foundation and at Oakland SPCA.

WORK HISTORY

1988–present	*Train Operator*	BAY AREA RAPID TRANSIT (BART) – Oakland
1993–95	***Owner/Manager***	**DOG'S BEST FRIEND**, pet care – Albany
1990–93	***Humane Educator*** (volunteer)	**LATHAM FOUNDATION and OAKLAND SPCA**
1985–89	*Owner/Operator*	COLLINS TRUCKING COMPANY– Albany
1982–84	***Kennel Aide***	**BERKELEY/EAST BAY HUMANE SOCIETY**

EDUCATION

A.S. Degree, Biology – Laney College, Oakland

ADMINISTRATION

GELIA THORNTON

6788 Virginia Street ◆ Oakland, CA 94602 ◆ (510) 990-3255

Objective: Position as purchasing and inventory control manager.

SUMMARY

- 10 years' experience in purchasing and inventory control.
- Successfully managed several automotive parts departments.
- Proven ability to transform a chaotic department into a smoothly running, efficient operation.
- Skill in maintaining optimum stock levels within budget allowances.
- Work effectively with a wide variety of people.

EMPLOYMENT HISTORY

1990–present **Senior Parts Technician** SHORELINE SHIPPING, Inc.
International cargo shippers – Richmond

- Requested bids from vendors for awarding of annual purchasing contracts; compared bids submitted, considering price, quality and shipping terms.
- Determined minimum and maximum stock levels for 6,000 line items.
- Conducted on-going reviews and updates of stock level figures, considering:
  ...expected and actual usage figures ...seasonal fluctuations
  ...planned equipment phase-out and replacement.
- Monitored invoices for accuracy of account coding; tracked expenditures for comparison with budgeted figures.

1989 **Foreman, Shipping/Receiving** ARMICO Inc., Wholesale/Retail Auto Parts

- Supervised and streamlined shipping/receiving department:
  – Identified and corrected the operational problems undermining efficiency;
  – Greatly increased the number of orders shipped daily;
  – Minimized the backlog of open orders.

1987–88 **Administrative Assistant** DREISBACH EXPORT PACKING – Oakland

1984–86 **Manager, Parts** LONG BEACH MAZDA – Long Beach, NY

- Set up an efficient parts department for a new Mazda dealership, starting with an empty room, a very large stock order, and a packing list in Japanese.
- Developed a variety of check-and-balance systems to reduce errors in inventory record keeping; achieved high accuracy (well under 1% loss) on annual inventory.

EDUCATION

English major – STATE UNIVERSITY OF NEW YORK, Plattsburgh

ROBERTA SWAN

115 Versailles Ave. • El Granada, CA 94018 • (415) 224-4443

Job objective: Position in program management/administration with a social service agency.

SUMMARY

- Experience in program planning and coordination.
- Strong motivation to help others live life fully.
- Skilled in handling the public with professionalism and sensitivity.
- Enjoy organizing complex projects and following through to completion.
- Excellent written, verbal, and listening skills.
- Special talent for assessing and improving office systems.

RELEVANT EXPERIENCE

PROGRAM PLANNING & COORDINATION

- Established and implemented a system for assigning credentials to convention staff and guests at 1984 Democratic National Convention:
  – Negotiated with national committee chairperson to upgrade staff credentials;
  – Made group presentations to staff, explaining complex credentialing procedures.
- Analyzed office paperwork flow in an AV studio and devised more efficient procedures.
- Developed a system for maintaining accurate record of financial transactions for a nonprofit workshop sponsored by Elisabeth Kubler-Ross Center.

MANAGEMENT & TRAINING

- Assessed appropriate duties for a new position in billing department; trained new employee to monitor rental equipment location, price job orders, and process paperwork.
- Researched and documented correct procedure for hiring freelance union studio talent, reducing hiring expense and office chaos.
- Delegated daily job assignments to 20-25 employees, and resolved logistics problems relating to timely delivery, setup and operation of audio-visual equipment.

COMMUNITY RESOURCES

- Located wide range of services for out-of-town convention production company:
  ...costumes ...props ...caterers ...audio-visual equipment.
- Conducted extensive research to locate appropriate site for weekend workshops supporting teenagers and children in effectively handling profound life crises.

WORK HISTORY

Freelance Production Services

— Conventions, film/TV, communications, 1988–present —

Administrative Assistant	UNITED WAY PRODUCTIONS, El Granada
Workshop Coordinator	ELISABETH KUBLER-ROSS CENTER, SF
Asst. Production Manager	FM PRODUCTIONS, Demo. Natl. Convention, SF
Office Services/Trainer	McCUNE AUDIO VISUAL, San Francisco
Equipment Rental Coordinator	" "
Production Coordinator	CREATIVE ESTABLISHMENT (convention prod.)

EDUCATION

B.A., Sociology – University of Colorado
Legal Assistant Program, University of San Diego

ADMINISTRATION

Rose satisfies the employer's need for clarity in a functional format by identifying where activities happened.

ROSE ELLINGTON

2773 Middletown Avenue • Alameda, CA 94501 • (510)-366-8447

OBJECTIVE: **Position as Member Relations Manager for a large Health Maintenance Organization.**

PROFILE

- Extensive experience with patient advocacy.
- Supervisory background in agency and retail environments.
- Demonstrated competence in coordinating programs.
- Excellent mediator, moderator and facilitator.
- Inspire and support others to work at their highest level.
- Licensed Marriage and Family Counselor.

RELEVANT EXPERIENCE

Supervision

- Managed staff of 25 youth in McDonald's retail establishment.
- Supervised up to five assistants in SFSU career development program.
- Supervised peer counselor trainees at San Francisco State Career Center.
- Taught and supervised 12 SFSU student teachers.
- Supervised two employees in my own public relations business.

Patient Advocacy and Counseling

- Served as patient advocate for two years, at San Mateo Co. Mental Health.
- Mediated domestic conflicts as marriage and family counselor for nine years.
- Developed training presentations for medical students on patients' rights.
- Acted as counselor and grievance liaison at Mental Health Assoc.

Program Administration

- Managed my own businesses for eight years, as a private therapist and as a publicist.
- Developed and coordinated numerous training programs.
- Prepared reports for state and county agencies, and for local agency directors.
- Participated in hundreds of staff meetings and planning sessions.
- Coordinated statistical research and analysis for grant proposals.
- Developed extensive marketing experience as owner of a public relations business.

PROFESSIONAL EXPERIENCE

1992–present	**Publicist/owner**	Out There!, public relations firm, Alameda
1991–92	**Career Counselor**	Self-employed, Berkeley
1989–91	**Coordinator, Career Program**	Woman's Resource, San Rafael
1987–88	**Manager**	McDonald's restaurant, Mill Valley
1985–91	**Marriage & Family Counselor**	Self-employed, SF Bay Area
1982–85	Graduate school student	San Francisco State University (SFSU)
1979–82	**Teacher/trainer**	San Mateo Co. Mental Health Assoc., Parent-Child Communication Program, Daly City
1976–80	**Counselor/Grievance Liaison**	San Mateo Co. Mental Health Assoc., Daly City

EDUCATION & TRAINING

M.A. San Francisco State University – San Francisco, CA
Marriage, Family & Child Credential (MFCC)
Adult School Credential – communication skills

ADMINISTRATION

Office and Program Support Resumes

Anthony Gabrielle	Customer service, entry	72
Carol Parker	Bookstore clerk	74
Darlene Jacobson	Program assistant	75
Sylvene Piercy	Office support/project coordinator	76
Estelle Havens	Clerical/bookkeeping, part-time	77
Joy Holland	Administrative assistant	79
Maryanne Hain	Executive assistant, international trade	79
Stephen Parker	Mailroom assistant/handyman	80
Stephen Scott	Office support/customer service, entry	81

ANTHONY GABRIELLE

2399 Blake Street
Berkeley, CA 94704
(510) 313-6602

Objective: Entry level position in customer service or project support with a major international corporation.

SUMMARY

- Well versed in current issues of international economics, reinforced by field studies at University of Edinburgh.
- Proven organization, communication and problem solving skills.
- Special aptitude for integrating diverse concepts.
- Natural flair for generating creative, innovative ideas.
- Lifelong interest in researching geography and demography.

RELEVANT EXPERIENCE

Economics, Geography, Demography

- Completed extensive course work integrating international economics and politics:
  - International Trade & the EEC
  - Comparative Economic Systems
  - Politics of Global Resource Scarcity
  - US Foreign Policy
  - Middle East & Global Perspective
  - Urban Geography of Great Britain
  - Europe in Crisis (history 1914–1945)
  - Thesis Seminar in International Relations
  - Government and the Economy
  - Third World Politics
  - Soviet Foreign Policy
  - British Public Policy
  - World Military Policy
  - Comparative Political Systems
- Conducted 5-year independent research in geography and demography, acquiring a professional level of knowledge in both areas.

Research

- Conducted long term in-depth research on Western European security issues, using libraries, World Affairs Councils, international publications, the Hoover Institution, and governmental archives, as primary resources.
  Organized, documented and successfully defended a policy-oriented thesis.
- Broadened my understanding of current international issues in history and political economy through extensive independent research at the National Library of Scotland.

Creative Problem Solving

- Greatly improved the effectiveness of merchandise displays at The Emporium's department store by creating more attractive and space-efficient arrangements.
- Devised more cost-effective, safe, and timesaving methods of handling merchandise, for a fast-paced mail-order service.

– Continued –

Tony makes it clear that all his work history was concurrent with his university studies. He'll use this resume for Informational Interviewing to more sharply define his job target; then he will re-write his resume to fit the more focused job objective.

RELEVANT EXPERIENCE, *continued*

Responsibility

- Entrusted with daily delivery of highly sensitive and valuable materials (bank statements, deposits, credit cards, vitally important computer printouts), serving as courier between hospital and medical building; maintained spotless driving record.

Diplomacy

- Played key role in successful negotiation of an international security issue, as member of model UN Security Council, at UC Berkeley:
  – Drafted a compromise; – Presented it to the Council; – Rallied support;
  – Effectively persuaded reluctant parties to accept the solution.

Languages

- Conversational competence in Italian; course work in French and German.

WORK HISTORY, Part-time concurrent with full-time education:

1996–summer	Overseas study in Great Britain	
1995	***Warehouse & Delivery***	KELLY TEMPORARY SERVICE jobs
1994–winter	***Stock & Display***	EMPORIUM, SF, via Kelly Temp Services
1994–summer	***Packing/Shipping***	EMPORIUM, 10th St. Distr. Center, SF
1993	***Bus person/Maintenance***	HARRIS' RESTAURANT, San Mateo
1992–summer	***Delivery/Maintenance***	THE MEDICAL CORP., San Francisco

EDUCATION

B.A., Dec. 1996, International Affairs, Minor in History;
University of California, Berkeley

5 months' overseas study of political economy, Great Britain

– References available on request –

CAROL A. PARKER

2330 Baker Street
Berkeley, CA 94704
(510) 834-8425

OBJECTIVE: Position as Bookstore Clerk at Avenue Books.

SUMMARY OF QUALIFICATIONS

- Thorough knowledge of bookstore layout, shelving books, locating books through reference books and microfiche.
- Have inventoried, organized and shelved materials for retail store.
- Able to help customers in a professional, concerned manner.
- Familiar with operating a cash register.
- Experienced in ordering and receiving materials.

WORK HISTORY

1995–1996	*Clerical Worker*	WESTERN TEMPORARY SERVICE, Berkeley, CA
1994	*Senior Clerk*	PERSONNEL OFFICE, UNIVERSITY OF CALIFORNIA, Berkeley, CA
1992–1993	*Retail Clerk*	LACIS Antique Lace Store, Berkeley, CA
1987–1991	*Librarian's Assistant*	BURLING LIBRARY, GRINNELL COLLEGE, Grinnell, IA
1981–1987	*Librarian's Assistant*	TOWN SCHOOL FOR BOYS library, San Francisco, CA

EDUCATION

A.A., Wood Technology, 1991–92
Laney College, Oakland, CA

B.A., Spanish, 1991
Grinnell College, Grinnell, IA

DARLENE JACOBSON

1219 Colusa Avenue
Berkeley, CA 94707
(510) 891-6500 messages
(510) 351-2828 home

Objective: Position as program assistant/administrative assistant.

SUMMARY

- Able to take responsibility for implementing a project.
- Strong commitment to cooperative teamwork.
- Excellent skills in facilitation, communication, presentation.
- Readily transcend educational/cultural/language barriers.
- Effective teacher and innovative designer of learning projects.

RELEVANT EXPERIENCE

Program Implementation

- Developed and implemented art programs at an Oakland Montessori school:
  - Designed and presented student projects, coordinated with teaching themes;
  - Supervised three teenage teaching assistants, focusing on job skills and goals;
  - Served as liaison to parents; organized their participation in field trips.
- Taught English As a Second Language to adults.

Administrative Support

- Assisted owners to remodel building for use as a produce market and ice cream store, involving light construction, plumbing, moving, ordering, pricing, advertising.
- Served as teaching aide in pre-school and grade K-6 classrooms.
- Reported on the development of children's skills at weekly teachers' meetings.
- Facilitated meetings: political action committees, business collectives, school staff.

Public Relations

- Created and presented educational slide show on refuges for women victims of domestic violence:
  - Took photos – Wrote script
  - Organized speaking engagements – Facilitated public discussion
  - Provided referral information on community resources.
- Greeted, oriented and scheduled clients, in several office settings:
  ...doctor's office ...social service hotline ...massage studio ...answering service.

EMPLOYMENT HISTORY

1996–present	**Masseuse**	BEVERLEY MASSAGE STUDIO, Berkeley, CA
1994–95	**Assistant Teacher**	MONTESSORI CHILDREN'S COMMUNITY, Oakland, CA
1992–93	**Phone Operator**	CENTRAL ANSWERING SERVICE, Berkeley, CA
1991	**Asst. Manager**	PRODUCE STORE, BOONT BERRY FARM, Boonville, CA
1990	**Asst. Editor**	CAREER CENTER FOR WOMEN, Santa Cruz, CA
1989	**ESL Teacher/Tutor**	Freelance, Paris, France
1984–88	**Student**	UCSC (spent 1987–88 in France, field study)
1983–84	**Teaching Assistant**	BECKNOLL DAY SCHOOL, Santa Ana, CA

EDUCATION

B.A., Community Studies – University of California at Santa Cruz, 1988

SYLVENE PIERCY

390 Wayne Place
Oakland, CA 94606
(510) 178-6663

Objective: **Office or customer support position**
involving project coordinating and problem solving.

SUMMARY

- Highly reliable self-starter; can be counted on to complete assignments without supervision.
- Creative in cutting costs and solving problems.
- Work well in a busy office, handling a wide variety of tasks.
- Sincerely enjoy helping people.

RELATED ACCOMPLISHMENTS

Project Coordinating & Supervising

- Prepared conference rooms for training and administrative sessions:
  ...ordered audio-visual & catering equipment ...assembled supplies packets.
- Coordinated large office parties: –prepared guest list and menu
  –set schedule –ordered furniture and supplies –coordinated set-up/clean-up.
- Oriented new employees; allocated jobs and monitored distribution of work load among clerks working on drafting/engineering team.

Problem Solving/Cost Saving

- Designed efficient reporting form and desk procedure for an engineering firm.
- Researched and found alternative source for typewriter needed by president's secretary, saving the cost of a new purchase.
- Submitted cost-saving ideas implemented by Bechtel Corporation, such as:
  – Proposed reclaiming/recycling desk supplies abandoned by relocated staff;
  – Designed more efficient time sheet which included space for overtime data.

Technical Skill

- Experienced using: –Microsoft Word –Microsoft Excel
- Performed light bookkeeping; monitored records:
  ...reviewed invoices for accuracy ...processed invoices and time sheets.
- Completed training in "Introduction to Computers" course.

Customer & Client Services

- Advised customers in selecting and installing ceramic tile.
- Advised employers on choice of fabric, colors, and styles for employee uniforms.

WORK HISTORY

1995–present	**Student,** business courses	Laney College – Oakland
1994–95	**Data Processing**	EQUITEC Financial Group – Oakland
1992–94	**Work Order Clerk**	BECHTEL CORP. – San Francisco
1989–92	**Manager** (Weekends)	CERAMIC TILE SALES – Oakland
1984–88	**Support Group Leader**	BECHTEL CORP. – San Francisco

Estelle is a high school graduate just entering college.

ESTELLE HAVENS
3744 35th Avenue
Oakland, CA 94605
(510) 443-9090

Objective: **Part-time clerical position in a bank or other business, assisting management with bookkeeping, filing, and light typing.**

SUMMARY OF QUALIFICATIONS

- Dependable; can be counted on to get the job done.
- Committed to a career in office work; motivated to learn and grow in responsibility and business skills.
- Good with figures; experience in bookkeeping and general office support.
- Excellent references from past employers and teachers.
- Well groomed; get along well with others.

RELATED SKILLS & EXPERIENCE

General Office Support

- Assisted store manager in orienting and assigning employees:
  - – Prepared new employee personnel folders; called in substitutes as needed;
  - – Monitored minors' work permits to insure they were still valid;
  - – Filed personnel records and managers' test results.
- Posted and filed official documents.
- Typed correspondence; answered phone; scheduled interviews; made reservations.
- Assisted local author in assembling material for new book:
  - – Contacted over 100 persons by phone and letter, successfully getting their permission to use personal data in new publication;
  - – Kept detailed records of results of contacts; updated client files;
  - – Mailed brochures and review copies of author's book;
  - – Accurately prepared subject cards for cross-indexing new book.
- Assisted candidate running successfully for office of County Supervisor:
  - – Called registered voters, describing the candidate's position on political issues.

Bookkeeping

- Accurately completed bookkeeping tasks at McDonald's in half the usual time.
- Recorded daily sales: – Tallied total items sold and computed total daily revenues;
  - – Recorded totals of wasted food and paper products;
  - – Audited the cash register records for each employee and reported errors.
- Earned an Outstanding Achievement raise at McDonald's for consistently accurate money handling and good relationships with customers.
- Assisted in computing employee hours and verifying accuracy of vendor statements.
- Balanced family checkbook statements and paid bills.

EMPLOYMENT HISTORY

1995–96 (summers)	**Office Asst.**	INCREDIBLE PUBLISHING, Berkeley
1993–94	**Bookkeeper**	McDONALD'S, Oakland
1992–93	**Cashier**	McDONALD'S, Oakland & Hayward
1992 (summer)	**Clerk**	HAVENSCOURT COMMUNITY CHURCH, Oakland
1990 (weekends)	**Cashier/Sales Asst.**	CERAMIC TILE CO., Oakland

EDUCATION & TRAINING

Freshman, CHABOT COLLEGE, Majoring in Entrepreneurship & Accounting
Completed courses in: Accounting, Law, Typing, Journalism

Joy maximizes the impact of her business training by spelling out the relevant courses she took.

JOY HOLLAND
660 Pinebluff Avenue
Vallejo, CA 94590
(707) 888-2557

Objective: Project Coordinator or Administrative Assistant position, involving personnel services, training, supervision, and/or payroll and bookkeeping.

PROFILE

- Demonstrated skill in supervising an efficient, well-run department.
- Extensive experience in personnel and office administration services.
- Enthusiastic, personable; professional in appearance and manner.
- Reputation for dependability and credibility.
- Take pride in doing a good job and achieving results.

RELEVANT EXPERIENCE

Project Coordinating
- Oversaw start-up of a new unit, Executive Credit Card, at Wells Fargo Bank:
  - Interviewed and hired employees to develop a staff of 10;
  - Set up initial Accounts Receivable ledgers for monthly posting;
  - Conducted credit investigations and developed profiles on new credit card applicants.

Personnel, Payroll & Bookkeeping
- Served as Group Benefits Coordinator at California Rubber and General Electric
- Processed Accounts Payable and Receivable; reconciled bank accounts.
- Processed payroll: calculated hours, taxes and deductions, and prepared special checks.

Training/Supervising
- Trained new employees at Wells Fargo Bank in computerized bookkeeping and use of office equipment and word processors.
- Trained employees of California Rubber on use of bookkeeping software.
- Supervised 10 clerical employees in the Accounting unit at Wells Fargo Bank.

EMPLOYMENT HISTORY

1988–present	***Bookkeeper/Admin. Asst.***	CALIFORNIA RUBBER MFG. CO., Vallejo
1986–88	***Financial Representative***	GENERAL ELECTRIC CO., Oakland
1977–85	***Credit Asst./Unit Supv.***	CROCKER BANK; WELLS FARGO BANK, SF

EDUCATION & TRAINING

Heald Business College, Laney College, College of Alameda

Business Coursework:
- Small Business Management
- Introduction to Business Law
- Principles of Accounting
- Business Math
- Word Processing
- Business English
- Payroll Accounting
- Income Tax Accounting

– References available upon request –

OFFICE SUPPORT

MARYANNE HAIN

790 Martinez Avenue, San Francisco, CA 94127
(415) 867-3522 • (415) 733-9097

Objective: **Executive Assistant in an international organization focusing on trade and business development in the Pacific Rim countries.**

PROFILE

- Traveled extensively, and often independently, in mainland China.
- Studied Mandarin Chinese intensively in Hong Kong and Taiwan.
- Keen awareness of regional politics; strong analytical skills.
- Readily transcend cultural and language differences.
- Young, enthusiastic, committed to professional growth.
- Liberal arts background, with high academic achievement.

RELEVANT EXPERIENCE

Knowledge of Chinese Customs and Culture

- Lived with local family in Taiwan, speaking Chinese exclusively in the home.
- Lived and studied with Chinese students in Hong Kong:
  - Participated in The International Club, encouraging cultural exchange and social interaction, as cultural liaison between Americans and Chinese students;
  - Studied Chinese international relations and Chinese literature;
  - Completed supervised historical research on relations between China and the West.

Knowledge of East Asian Politics and Economy

- Attended extra-curricular events sponsored by Yale-China Association, e.g.:
  ...speeches by diplomats; ...political/social events on the future of Hong Kong.
- Participated in a series of conferences analyzing the difference between Chinese and Western law, impacting the transition of Hong Kong to Chinese control.
- Tutored Vietnamese refugees in English; greatly increased my political awareness.

Management Support

- Worked up to increasing levels of responsibility with Hotel Group of America, starting out as desk clerk at Hotel Union Square:
- Trained newer desk clerks in computer use, guest relations, billing and reservations.

WORK HISTORY

current	**Head Desk Clerk***	DUBUQUE HOTEL, San Francisco
1994–95	**Desk Clerk***	HOTEL UNION SQUARE, SF
1993–94	**Cashier***	VIE DE FRANCE Restaurant, SF
1993 summer	**Cashier**	SOURDOUGH PUFF CO., bakery, SF
1992 summer	**Office Assistant**	LEOUNOUDAKIS & FORAN law firm, SF

*full-time summers, half-time during school year

EDUCATION

B.A., History – UNIVERSITY OF CALIFORNIA, BERKELEY
1995–96 study abroad: CHINESE UNIVERSITY OF HONG KONG and YALE-CHINA LANGUAGE CENTER
Summer 1995, National Taiwan Normal University, Mandarin Training Center

STEPHEN P. PARKER
P.O. Box 3289
Berkeley, CA 94703
(510) 658-9229

Objective: Maintenance Handyman/Mailroom Assistant for the Nature Company headquarters.

PROFILE

- Self-motivated; able to learn anything on my own initiative.
- Excellent record of dependability and reliability.
- Lifetime interest in nature and nature studies.
- Wide range of manual skills.

WORK HISTORY

1990–present ***Nature Photographer*** – Freelance, part-time; NY & West Coast
- Published photos in national and regional magazines; agency-represented.

1994–96 ***Maintenance Worker*** – SOUTHSIDE MALL, Oneonta, NY
- Repaired and maintained plumbing, door hardware, grounds-keeping equipment;
- Light carpentry as needed;
- General cleaning, building security, opening and lock-up.

1993 ***Bicycle Mechanic*** – ALL-AMERICAN SPORT SHOP, Oneonta, NY
- Repaired and maintained all types of bicycles;
- Fabricated parts as needed.

1989–92 ***Auto Mechanic*** – VAN'S AUTO SERVICE, Oneonta, NY
- Repaired and maintained all makes of automobiles and light trucks;
- Developed expertise in brakes, suspension, exhaust, tune-ups, tire repair, mounting and balancing.

1985–89 ***Dept. Mgr./Stock Clerk*** – GREAT AMERICAN FOOD STORES, Cooperstown, NY
- Ordered and rotated stock; generally supervised dairy department.

Stephen recently graduated from high school and has little paid experience, so he makes the most of it by documenting his relevant skills.

STEPHEN M. SCOTT

9065 Hillegass Avenue • Berkeley, CA 94705 • (510) 898-3099

Objective: Entry level position in office support or customer service.

SUMMARY

- Excellent telephone communication skills.
- Friendly, courteous and articulate.
- Take pride in doing a good job; willing to learn.
- Familiar with computer database and word processing.
- General working knowledge of business machines.

RELEVANT EXPERIENCE

Office

- Typed letters, envelopes, labels and invoices for student publications office.
- Ran copies on Xerox machine; refilled paper and toner.
- Waxed camera-ready copy in preparation for layout of newspaper.
- Completed basic accounting class; familiar with use of calculator.
- Answered phones, sold computer training courses, and enrolled students in IBM-PC courses as Administrative Assistant at Computers!Computers! retail store.
- Assisted architectural space planning project at City Hall, measuring and recording furniture and work areas on each floor and in each department.

Telephone Skills

- Filled in as receptionist at Center for Independent Living, handling a heavy load of incoming calls and relaying messages to staff.
- Made over 100 PR calls to businesses for ADMARK Corp.
- Took incoming calls for campus newspaper staff at University of Houston.

Computer Knowledge

- Completed introductory training courses at Computers!Computers!, covering: ...disk operating system ...database ...WordPerfect.
- Input 1000 names and addresses into Filemaker database for ADMARK Corp.

WORK EXPERIENCE

1996 (temp, 1 mo.)	*Office Assistant*	COMPUTERS!COMPUTERS!, SAN FRANCISCO
1995 (part-time)	*Office Assistant*	Short-term jobs (up to 2 months), assisting in various PR assignments for ADMARK Corp. and CENTER FOR INDEPENDENT LIVING, Berkeley
"	*Carpentry Asst.*	HAL HOWARD FLOOR REFINISHING, Berkeley
1995 (summer)	*Inventory Clerk*	BERKELEY CITY HALL, Health & Human Svcs. Dept.
1990–94*	*Office Assistant*	STUDENT PUBLICATIONS office*, Univ. of Houston

*summer jobs assisting in my father's newspaper publication office

EDUCATION

Completed High School – Bellaire Senior High, Bellaire, TX, 1996

OFFICE SUPPORT

Finance and Accounting Resumes

Munana Fehreshta	Accounting, entry level	83
Polly Kelsa	Accounting/financial consulting	84
Suzanne Chew	Auditing, entry level	85
Sonia Morena	Accounts receivable manager	86
Mack Anderson	Management trainee, accounting	87
Andrea Graham	Auditing, entry level	88
Claudia Peterson	Service rep, accounting software	89
David Douglas	Financial analyst	90
Hannah Cortland	Bookkeeper/receptionist	91
Katherine Lawrence	Bookkeeper, full-charge	92
Lorna Burlingame	Financial consultant, investments	93
Lynne Joussart	Business management/financial planning	94
YonSoon Kwang	Junior accountant	95

MS. MUNANA FEHRESHTA

(510) 771-1212 home
(510) 773-6984 work

335 Turling Boulevard
El Cerrito, CA 94530

Objective: Entry level staff position in a public accounting firm.

SUMMARY

- Experience in auditing, payroll, and preparation of corporate taxes.
- Passed all four parts of the CPA examination at first sitting.
- Excellent team worker; function well under pressure.
- Deeply committed, professional attitude.
- Managed a retail coffee business, turning a $10,000 loss into a $20,000 net profit; set up a second business which was immediately profitable.

RELEVANT EXPERIENCE

1996–present ***Junior Accountant*** TOM DREYER, CPA – Pleasant Hill, CA

Auditing
- Conducted a review of lending institution's records:
  - Selected random sample of transactions;
  - Conducted compliance testing;
  - Drafted final report for the client.

Accounting & Taxes
- Prepared quarterly payroll and sales taxes for over 50 individuals.
- Filed corporate taxes for three corporations.
- Reconciled bank statements to clients' books.
- Reconstructed accounting records from clients' checks and cash receipts.
- Searched out-of-state tax codes relevant to client's income tax.

Computer Skills
- Posted clients' journals to computer, made adjusting journal entries, and created general ledgers.
- Generated financial statements and accountant compilation and review notes.
- Worked with Macintosh, DOS and Windows operating systems.
- Currently customizing a Multiplan program for two small businesses.

1992–95 ***Store Manager*** EL CERRITO COFFEE MILL (retail coffee) and concurrently TOM SHAMSI SERVICE (office machines servicing)

- Advised my employers on:
  - Cost-advantageous timing of major equipment purchases;
  - Cash flow alternatives and comparative sources of investment capital;
  - Options for minimizing federal tax.

EDUCATION

B.S., Accounting, 1995 – ARMSTRONG UNIVERSITY, Berkeley

POLLY KELSA

59912 Le Conte • Berkeley, CA 94709 • (510) 987-6543

OBJECTIVE: Accounting position, including financial consultation and training.

SUMMARY

- Four years' experience in bookkeeping for small businesses and corporations.
- Enjoy consulting with clients, and helping them get what they want.
- Commitment to professional growth and development in financial services.
- Outstanding talent for assessing clients' needs and developing individualized financial systems.
- Extremely dependable in completing projects accurately and on time.

RELEVANT EXPERIENCE

Bookkeeping & Accounting

- Served as full-charge bookkeeper and financial manager for several businesses, responsible for:
  ...general ledger ...cash disbursements journal ...cash receipts journal
  ...payroll and payroll taxes ...accounts payable ...accounts receivable
  ...bank reconciliation ...budgeting ...financial statements.
- Consulted with CPAs on behalf of businesses, presenting all financial materials to prepare for end-of-year taxes.

Needs Assessment/Advising

- Successfully advised and counseled small business clients on financial strategy, employing effective counseling methods. Combined candid assessment of current status with proposals for improving profitability.

Organization/Administration

- Trained and supervised novice bookkeepers in bookkeeping skills.
- Advised on and implemented start-up procedures for small businesses in the State of California, involving business license, bank account, fictitious name statement, resale number, state and federal employer ID number.
- Served as financial officer on committees and boards of several organizations.
- Established my own financial bookkeeping service.

EMPLOYMENT HISTORY

1995–present	**Owner/Consultant**	BUSINESS STRATEGIES, Berkeley, CA (serving 8 small business clients)
1993–96	**Co-Owner/Financial Mgr.**	BEAUTY SECRETS, Berkeley, CA
1992–95	**Financial Manager**	MIRO CHIROPRACTIC, Berkeley, CA
1989–91	**Bookkeeper**	JANET ROGERS, CHIROPRACTOR, Berkeley, CA
1988–89	**Bookkeeper**	BARNES TRAVEL AGENCY, Oakland, CA

EDUCATION & TRAINING

B.A., Communication Arts & Sciences – QUEENS COLLEGE, NY, NY 1983
Class in General Accounting – VISTA COLLEGE, Berkeley, CA 1996
Class in Psychology - JFK University, Orinda, CA

As a new graduate, Suzanne's education counts heavily in her qualifications so it is placed near the top of her resume.

SUZANNE CHEW

Campus Address
2230 Haste St., Apt. 999 • Berkeley, CA 94704
(510) 523-7867

Alternate Address
226 Calamari Ct. • Concord, CA 94521
(510) 907-0742

Objective: Entry level position in Audit Department.

PROFILE

- Bachelor of Science in accounting.
- Successful in mastering accounting theory and technical skills.
- Dedicated to professionalism, highly motivated toward goal achievement.
- Experience in coordinating projects involving people and activities.

EDUCATION

B.S., Accounting – University of California, Berkeley, May 1996
Honor student since Fall semester 1995
Accounting G.P.A. 4.0 – Overall G.P.A. 3.685

Affiliations: U.C. Berkeley Honor Society; Professional Women's Association; Undergraduate Business Association

EXPERIENCE & SKILLS

Accounting Knowledge
Developed solid theoretical base in financial accounting at UC Berkeley Business School:
- Able to set up balance sheets and income statements, and analyze clients' assets and liabilities.
- Studied laws relevant to accounting and other business applications.
- Currently advise students on problems in accounting classes and grade their homework.

Business Project Coordination
Coordinated focus group studies for Walnut Creek market research firm:
- Invited prospective group members by phone; provided participants with study materials.
- Solicited phone interviews from random samples, consistently convincing participants of the legitimacy of the project and the importance of their opinions.
- Collaborated with co-workers to assure consistent coding of research materials.
- Edited market research interviews; entered coded data into computer, and generated reports.
- Gave oral evaluations of market research interviews to clients from ad agencies.
- Earned the confidence of the firm's manager, and was invited to assume more responsibility through a supervisory position.

WORK HISTORY

1995 fall	**Reader**	U.C. BERKELEY BUSINESS SCHOOL – Berkeley, CA
1994 summer	**Sales Clerk**	SHOE CITY – Cornwall, CA
1993–1994	**Interviewer**	PDQ MARKETING RESEARCH – Walnut Creek, CA
1991–1993	**Interviewer**	WESTERN QUICK SEARCH – Cornwall, CA

Sonia describes the US equivalent of her Guatemalan degree.

SONIA MORENA

69 Sixth Avenue
San Francisco, CA 94109
(415) 411-2675

OBJECTIVE: Position managing an Accounts Receivable/Payable Department.

SUMMARY

- Over 10 years' experience in Accounting.
- Can adapt immediately to any accounting system.
- Talent for resolving business and accounting problems.
- Ability to work independently.
- Competent with computer applications for wordprocessing and accounting.

PROFESSIONAL EXPERIENCE

1995–present **Accountant** (temp) REED ADVERTISING, SF

- Made transitions from hand-operated systems to computerized systems, and from one computer system to another.
- Established cooperative relationships with institutions (banks, credit associations), to get credit information quickly.

1980–95 **Accountant** KXYZ RADIO, San Francisco
KBHK-TV44, San Francisco
KSFO RADIO, San Francisco

- Set up an initial accounting system for three small businesses and trained their staff to maintain it.
- Developed competence in all accounting aspects:

–accounts payable	–personnel	–credit management
–accounts receivable	–payroll	–expense reports audit
–general ledger	–sales commissions	
–financial statements	–capital accounts (assets).	

1979–80 **Accts. Receivable** SWANK-BECHELLI LEATHER WHOLESALER, San Francisco

- Reconstructed accounting records lost in a fire, analyzing and reorganizing materials from vendors' files in several different states; the company was able to recover all the monies due.
- Reduced bad debts from 33% to 1.5% by finding ways clients could clear up old debts and reinstate their credit with the company.

EDUCATION

B.S. equivalent, Business Administration;
Escuela de Comercio, Guatemala

Postgraduate study: Accounting – Boston University

Bilingual in Spanish & English; U.S. citizen

MACK ANDERSON
Post Office Box 8554
Emeryville, CA 94608
(510) 845-8344

Job objective: Entry level Accounting or Management Trainee

SUMMARY OF QUALIFICATIONS

- Completed requirements for a B.S. in Business Management; minor in Computer Science, Communication, and Finance.
- Eligible for CPA exam.

RELEVANT EXPERIENCE

EDUCATION

- Total of 300 semester hours of undergrad work completed. Coursework included:

-Business Management	-Cost Accounting
-Intermediate Accounting	-Taxation
-Personnel Management	-Money and Banking
-Financial Management	-Macro/Micro Economics
-Business Machines	-Intermediate Calculus I & II
-Production Management I & II	-Introduction to Marketing

-Computer Science (dBase, Lotus, Word Processing, FORTRAN, UNIX, COBOL, PASCAL, Computer Math, Computer Architect, CAD)

ACCOUNTING RELATED

- Kept Accounts Payable/Accounts Receivable ledger for Air Force food service unit.
- As liaison between Naval Supply Center and customers contracted for supplies, compared price and quality of various products from suppliers.
- Purchased supplies for Hoover Dam engineers; served as courier/liaison between regional director and other senior executives.

WORK & EDUCATION SUMMARY

1985–88	General Clerk, Hoover Dam/Bureau of Reclamation, Boulder City NV
1989–93	Part-time Student, Cal State Hayward and City College of San Francisco
1991–93	Supply Clerk, Oakland Naval Supply Center
1993–95	Part-time Night Watchman
1996	Odd jobs while seeking full-time employment

Mack was a homeless job hunter with a "ragged "recent work history. We reversed the usual order of his job history to play down his recent unemployment and inserted "part-time student" to help fill an awkward gap.

ANDREA GRAHAM
333 San Jose Ave.
San Francisco, CA 94110
(415) 689-2468

Objective: Entry level auditing position with a national CPA firm.

SUMMARY

- Ten years' experience in accounting and taxation.
- Highly successful in establishing, streamlining and automating accounting systems.
- Strength in recognizing, analyzing and solving problems.
- Effective in a dynamic and challenging environment.

RELEVANT EXPERIENCE

Assessment of Financial Records

- Evaluated financial statements for a wide range of clients, both as a staff accountant in a CPA firm, and as a self-employed bookkeeper:
  - Conducted thorough examination of clients' financial source documents to assure that proper accounting treatment was applied, and that records were complete;
  - Investigated account balances to verify their accuracy and actual existence;
  - Generated adjusting journal entries to provide accurate financial statements.

Accounting Systems Design & Implementation

- Assessed the efficiency of small business internal accounting systems, implementing procedures to improve financial accuracy and operating cost-effectiveness; this involved: reducing staffing, automating manual records, improving cash collections, training personnel and developing budgets.

Client Relations/Communication

- Interpreted financial statements for clients, explaining variances in budget analyses and discrepancies in annual comparative statements.
- Maintained excellent client relationships, securing trust and confidence by providing complete, accurate and timely financial services.

EMPLOYMENT HISTORY

1994–present	***Accountant***	LARSONN MANAGEMENT CO., SF
1991–94	***Self-employed Accountant***	A. GRAHAM ACCOUNTING SVCS., SF & SB
1990–93	***Staff Accountant***	RICHARD A. BERTI, CPA, Santa Barbara
1989–90	***Full Charge Bookkeeper***	INTERNAT'L. TRANSDUCER CORP., "
1986–89	***Staff Accountant***	GUNTERMANN, BALL et al, CPAs, "

EDUCATION

B.A., Accounting – GOLDEN GATE UNIVERSITY, SF (3.63 GPA in Accounting)

"Maternity leave" and "family management" present her past priorities and leave no gaps.

CLAUDIA PETERSON

577 Steiner Street
San Francisco, CA 94117
(415) 911-0808

Objective: Position as service representative for accounting software firm, specializing in conversions.

SUMMARY

- Excellent teacher / trainer; patient and effective with a wide range of personalities.
- Successful in identifying and solving computer related problems.
- Project oriented, sticking to a task until completed.
- Sharp in learning and comprehending new systems and methods.
- College level training in accounting, with 3.8 GPA.

RELEVANT EXPERIENCE

Bookkeeping

- Reconciled loan payment records between servicing company and 100 lending institutions.
- Reconciled cash records to computer records for over 100 accounts on a monthly basis.
- Prepared monthly payroll, paid bills and processed tuition payments for private preschool.

Teaching/Supervising

- Trained eight people in investor accounting, most of whom had no previous experience.
- Wrote a step-by-step instruction manual on the daily, weekly and monthly reporting requirements for 50 investors, minimizing training time for new employees.
- Maintained cordial working relations, while explaining and clarifying others' clerical errors.
- Interviewed and hired four clerical staff members.

Computer Usage

- Learned and mastered accounting clerk job and computer use in only two weeks.
- Worked with computer analyst in development of computerized specialty reports.
- Assisted in implementation of new program on a PC for accounts payable.
- Input monthly account records on a PC and generated trial balance.

Problem Solving

- Successfully reconstructed investor accountant job from evidence found in desk and files, although no one remaining with the company had any knowledge of this position.
- Balanced seven months of critical reports for a loan servicing company, involving 350,000 loans in their six major accounts, which had been neglected for five months.
- Designed an account coding system to eliminate dual coding and to avoid confusion and time wasted correlating reports.
- Reviewed company procedures, identifying sources of high error frequency, and submitting recommendations to supervisors.

EMPLOYMENT HISTORY

1996–present	***Treasurer/Bookkeeper*** *part-time*	FIRST PRES. CHURCH PRE-SCHOOL, Newark
1995–96	*Family management*	
1993–94	***Investor Accountant***	FIRST DEED CORP, Walnut Creek
1988–93	*Maternity leave*	
1987–88	***Accts. Reconciliation*** *part-time*	BALDWIN & HOWELL, San Francisco
1984–87	***Investor Acc't./Supervisor***	MASON-McDUFFIE CO., Berkeley

EDUCATION

Accounting & Business, ARMSTRONG COLLEGE, Berkeley; SIERRA COLLEGE, Rocklin
– Selected as Bank of America Business Award student at Sierra College

David lists each position at the hospital, showing the range of job titles.

DAVID DOUGLAS

245 Central Avenue • San Carlos, CA 94070 • work: (415) 372-2000

Objective: Position as financial analyst, senior accountant, assistant controller, or cost accountant.

Profile

- ◆ 10 years' experience in accounting with strong academic background.
- ◆ Strong analytic and problem solving abilities.
- ◆ Thorough and well organized in completing projects.
- ◆ Ready and willing to assume responsibility.
- ◆ Committed to professionalism.

PROFESSIONAL EXPERIENCE

Accounting

- Researched and produced complex monthly financial statements.
- Documented restricted funds to clarify amounts available; also responsible for distribution.
- Developed and upgraded shared data processing system:
  - Improved the accessibility of financial and statistical data for reporting purposes;
  - Allowed easier access to project information for non-accounting managers.
- Used "Lotus 1-2-3" for a wide variety of applications.

Project Management/Supervision

- Successfully filled in for accounting manager on short notice:
  - Assembled all work papers for year-end audit and handled auditors' inquiries;
  - Supervised payroll, accounts payable, data processing, cashiering & financial statements.
- Served as Assistant Supervisor in accounts receivable department.

Financial Analysis

- Implemented a new program to develop accurate costs for hospital diagnostic categories, enabling the hospital to develop an equitable reimbursement scheme for contract negotiations:
  - Interviewed department heads to establish financial value of procedures;
  - Coordinated statistical and financial data from various sources;
  - Designed appropriate spreadsheet format, using Excel on a PC.

EMPLOYMENT HISTORY

1990–present	**Cost Analyst**	MIDDLETON HOSPITAL, Middleton, CA
1988–90	**Financial Analyst**	" "
1987–88	**Accountant**	" "
1986–87	**Budget Analyst**	" "
1985–86	**Accounts Receivable Supervisor**	" "

EDUCATION & PROFESSIONAL AFFILIATION

B.A., Economics – San Francisco State University, 1984
Additional graduate work and classes in Accounting
Member - Board of Directors, National Association of Accountants

HANNAH CORTLAND

1990 Grand Avenue
Berkeley, CA 94703
(510) 220-1990

OBJECTIVE

Bookkeeper / receptionist position, with emphasis on accounts receivable, office coordination, scheduling, computer data entry.

SUMMARY

- Exceptionally responsible, diligent and thorough.
- Fast learner with a wide range of practical skills.
- Special talent for office organization.
- Excellent verbal and written communication skills.
- Able to handle challenging tasks in a busy office.

EXPERIENCE

Bookkeeping

- Computed and prepared monthly billings for over 100 employment agency clients, achieving a record of exceptional accuracy.
- Maintained records of daily income, and prepared agency's bank deposits.
- Calculated payroll deductions: state / federal taxes, disability, social security.
- Developed monthly financial report for Board of Directors of Athena House.

Office Coordination/Scheduling

- Created an efficient filing system for Here's Help employment agency, transforming haphazard records into readily retrievable form.
- Coordinated wide range of logistics for office functions:
  - Supervised repairs of office equipment; ordered supplies;
  - Performed minor repairs and maintenance;
  - Researched sources / selected new office equipment.
- Scheduled screening interviews of job applicants; filled in for office manager.
- Responded by mail to employer requests for insurance-related information; devised form letter for responding to inquiries from potential job applicants.
- Designed and produced promotional brochure for employment agency:
  – Edited text; – Arranged for typesetting; – Laid out graphics.

Computer Data Entry

- Accurately entered personnel data for over 1000 applicants, using customized computer program; updated and maintained each applicant's records.
- Taught myself computerized word processing and basic spread sheet, and taught other employees to use business programs.

WORK HISTORY

1995–present	**Receptionist/Bookkeeper**	HERE'S HELP, INC. temp. agency
1990–94	**Housekeeper,** Self-employed	East Bay clients
1988–90	**Bookkeeping trainee**	ATHENA HOUSE residential treatment, Santa Rosa
1987	**Accts. Payable/Cashier**	COMMUNITY MARKET COOPERATIVE, Santa Rosa
1986	**Produce Buyer**	OUR SMALL PLANET restaurant, Santa Rosa

EDUCATION

Sonoma State University, 1984–87 Interdisciplinary Studies

KATHERINE LAWRENCE

1490 Acton Street
Berkeley, CA 94702
(510) 488-1546

OBJECTIVE: Position as Full-charge Bookkeeper.

SUMMARY OF QUALIFICATIONS

- Five years' accounting experience, including automated general ledger.
- BA degree plus 2 years' education in Basic Accounting.
- Thoroughly familiar with all general accounting procedures.
- Highly valued and effective supervisor as well as co-worker.
- Can be counted on to get the job done.

EMPLOYMENT HISTORY

1990–present **Senior Acctg. Clerk** SOUTHLAND CORP., Corte Madera, CA

- Mastered a variety of accounting systems such as owner's draw system, merchandise invoice system, report-on-demand system, shared expenditure schedule, and individual store inventory audits.
- Operated an automated accounts payable system covering approximately 120 accounts, paid weekly, bi-weekly and monthly.
- Supervised 3 junior accounting clerks and audited their accuracy.
- Trained 12 junior accounting clerks and 2 managerial trainees in retail accounting.
- Posted to a general ledger and generated monthly and annual reports.
- Reconciled bank statements, involving over $1 million per month.

1989–1990 **Payroll Clerk** LAWLEY MANUFACTURING CO., Emeryville, CA

- Converted a manual payroll system to a sophisticated computerized system, researching and organizing documents back 7 years.
- Monitored for full data accuracy in the automated payroll system covering 32 retail outlets, involving 60 factory workers in four unions.

1988–1989 **Bookkeeper** STAINED GLASS GARDEN INC., Berkeley, CA

- Accomplished all accounting activities for this small retail glass business, including financial statements, accounts payable, accounts receivable and payroll.
- Maintained materials inventory and handled purchasing for a retail outlet.
- Analyzed financial statements, produced reports, input data on an online computer system.

1987–1988 **Inventory/Sales Clerk** NEW RENAISSANCE GLASS WORKS, Oakland

1986–1987 **Teaching Assistant** NAPA COLLEGE, Napa, CA

- Organized and taught classes, including development of class materials; assisted in teaching an elementary accounting course.

1983–1985 **Supply Clerk** US ARMY, Monterey, CA

EDUCATION & TRAINING

B.A., Linguistics, 1984 – University of Michigan
Accounting Program, Napa College • Programming, University of California

LORNA BURLINGAME
1835-A Broderick St.
San Francisco, CA 94115
(415) 122-6754

Objective: Position as financial consultant with an investment advisory company.

PROFILE

- Thrive on challenge and new opportunities for accomplishment and success in helping others achieve their objectives.
- Readily inspire the confidence and trust of clients.
- Extensive knowledge of financial instruments.
- Sharp analytic, problem solving and presentation skills.
- Legal background and training.

RELEVANT EXPERIENCE

FINANCIAL PLANNING

- Developed customized financial plans for Merrill Lynch clients:
  - Clarified their most immediate objectives and longer term goals;
  - Assessed their financial resources and tolerance for risk;
  - Explored various investment options and structured plans consistent with their experience, obligations, resources and risk temperament;
  - Implemented and periodically updated financial plans, keeping clients appraised of the status of their investments.

RESEARCH, ANALYSIS & EVALUATION

- Researched and analyzed various investment instruments:
  ...stocks ...bonds ...convertible bonds ...mutual funds ... CDs ...insurance products.
- Researched market conditions affecting clients' current and future financial strategies.
- Developed a sharply defined, critical mode of evaluating clients' needs and analyzing investment instruments/strategies, as a result of legal experience and training.

PROMOTION/PRESENTATION

- Delivered presentations on various retirement plans and on mutual funds to professional associations and private investment groups.
- Generated a high level of referrals from previous investment clients.

EMPLOYMENT HISTORY

1994–present	*Attorney*	DAVIS & ROWAN, attys., San Francisco
1993–94	*Financial Consultant*	MULHADY, PRUITT & PRICE, attys., SF
1992–93	*Law Clerk*	PARKER, ACACIA, FLOTOW & HOWARD, attys., SF
1992	*Mediator/Arbitrator*	COMMUNITY DISPUTE SERVICES, SF
1991–92	*Judicial Extern*	SAN JOSE SUPERIOR COURT, San Jose

EDUCATION

J.D., UNIVERSITY OF SANTA CLARA LAW SCHOOL, 1992
B.A., Psychology, UNIVERSITY OF CALIFORNIA at SANTA CRUZ, 1988

LICENSES & AFFILIATIONS:
Series 7 Brokerage Securities License
State Bar of California, License & Membership • Amer. Bar Assn., Family Law Section
San Francisco Chamber of Commerce • Junior League of San Francisco

Lynn could have used a choronological format (since she's staying in the same field) but this format seemed to give her experience more coherence.

LYNNE JOUSSART

299 Oakland Avenue #102 • Oakland, CA 91610
(510) 453-2104

OBJECTIVE

A part-time position providing business management, financial planning and full-charge bookkeeping.

SUMMARY

- Strong grasp of accounting; experienced in all phases of accounting.
- 10 years' experience in the business world.
- Successful in translating long-range organizational objectives into effective financial plans.
- Set up four businesses and designed their bookkeeping systems.

PROFESSIONAL EXPERIENCE

FINANCIAL PLANNING

- Produced budgets and cash flow projections for several nonprofit organizations.
- Developed detailed financial plans for two businesses.
  – Assisted management in refining their corporate goals;
  – Generated the financial data to complete a business plan.

BOOKKEEPING & ACCOUNTING

- Performed full-charge bookkeeping through financial statements:
  – payroll and payroll taxes
  – accounts receivable & billing
  – accounts payable
  – cash receipts and disbursements
  – bank reconciliations
  – inventory control.
- Prepared corporate and property tax returns for three nonprofits.
- Generated financial statements for small businesses on a computer.

BUSINESS MANAGEMENT

- Set up the initial books for four new businesses, and filed their incorporation papers.
- Extensively analyzed financial statements for small corporations.
- Filed periodic financial reports to supporting foundations.

EMPLOYMENT HISTORY

1995–present	**Financial Consultant**	URSA INSTITUTE & affiliates, San Francisco (large nonprofit firm; business consulting / social research)
	Financial Consultant	NAT'L. ALLIANCE AGAINST VIOLENCE, Inc., Oakland
1992–1994	**Business Manager**	LIVE OAK INSTITUTE, Oakland (training programs in nursing homes)
1992–1994	**Bookkeeper,** Full-Charge	BERKEY & ASSOC., Berkeley (bookkeeping and consulting service)
1989–1991	**Bookkeeper**, Part-time & Consulting, for 5 Boston area firms	
1987–1989	**Grain Buyer**	NEFCO, Cambridge, MA (coop. regional warehouse)
1985–1986	**General Mgr./Founder**	BUFFALO MOUNTAIN COOP, Hardwick, VT

EDUCATION

Accounting – Northeastern University, 1989
Liberal Arts – Goddard College, 1985-86

Ms. YonSoon Kwang

9874 MacArthur Boulevard
Oakland, CA 94609
(510) 652-9999

Job Objective: Position as Junior Accountant

Summary of Qualifications

- On-going studies toward degree in Business / Accounting; courses completed: Financial Accounting, Managerial Accounting, Small Business Management
- Competent and experienced in using QuickBooks for Windows, and handling accounts payable and accounts receivable.
- Able to journalize transactions, categorizing accurately.
- Experience in handling computerized payroll, balancing daily transactions, and bank reconciliation.
- Hard worker; honest, loyal, and extremely reliable.
- Learn quickly; resourceful in resolving problems.
- Get along well with co-workers; always willing to help.
- Excellent relations with vendors, customers, bankers, and business associates.

RECENT WORK EXPERIENCE

OFFICE MANAGER / BOOKKEEPER — 1994–present
Claremont Auto — Albany, Berkeley, Hayward, Oakland
- Oversaw the bookkeeping for all four store locations.
- Produced monthly financial statements and bank reconciliations.
- Resolved customer complaints and personnel problems.

Oakland store (headquarters):
- Balanced daily transactions and verified that job costs were properly charged.
- Deposited daily receipts; processed accounts payable and accounts receivable.

MARKETING ASSISTANT to the President — 1992–94
Investor's Guides, Inc., Walnut Creek
- Authorized payments to vendors.
- Organized and supervised large-scale project mailings. Identified the most profit-generating mailing lists to use for each project.
- Tracked daily project sales to determine profit and loss.
- Analyzed sales data and wrote reports; developed future projects based on these results.

CUSTOMER SERVICE REP — 1991–92
Metropolitan Bank, Albany CA
- Assisted bank customers with transactions.
- Logged transactions into computer. • Balanced daily debits / credits.

Additional Earlier Experience

Programmed consumer price indexes for the Korean government.

EDUCATION

A.A. (in progress), Business Management/Accounting —Vista College, 1994–present
A.A. degree, Computer Science — Kwang-Won Electronics Univ., Seoul, Korea
English, 1994 — U.C. Berkeley Extension
Teller Training, 1992 — Teller Training Institute, Concord

References available on request.

Technical and Computer-related Resumes

Susan Pieper	Computer software support	97
Lydia Silvers	Applications engineer, plastics	98
Rochelle Normandy	Computer/PC support services	100
Coronet Gibson	Research services coordinator, computer firm	101
Norman Whitmore	Technical writing/scientific writing	102
Mark Ebrahimi	Telecommunications supervisor	103
John Bridges	Research associate/biotechnology	104
Larry Elton	Engineering management	105
Patricia Delich	Instructional design/online technologies	106
Michael Hayakawa	Imaging services director (radiologist)	108

SUSAN PIEPER

3680 Alcatraz Avenue
Oakland, CA 94602
(510) 777-1224

Objective: Position in software support for a computer or software retailer.

PROFILE

- Exceptional ability to quickly master new software and apply its full range of capabilities.
- Accurately interpret customers' problems and offer the best resolution.
- Outstanding telephone communications; patient, personable and receptive.
- Six years' experience with personal computers of all types.

RELEVANT EXPERIENCE

1990–present **Production Manager** SYSTEMS DEVELOPERS, Oakland, CA

Training
- Trained seven magazine publishers in computer application for their industry:
  - Introduced them to elementary computer use;
  - Taught them how to use Remote Copy Input program, designed by my company;
  - Advised on which parts of the program applied to their current needs;
  - Assisted them, on an on-going basis, in using increasingly more of the program's components.

Troubleshooting & Problem Solving
- Advised clients on what is technically possible or impossible and how to achieve particular printing effects.
- Successfully resolved timing problems between client, production and printer.
- Saved important photo-history data for a client by effectively backtracking to the source of the problem and correcting the original file.
- Reorganized user areas on customer's hard disk, restoring compatibility of file locations so the program could retrieve them.

Software Development & Program Application
- Assisted programmers in design of a program to translate computer generated ads into full-page compositions, by advising on how typesetting works.
- Identified specific applications for a newly written program.
- Tested new programs for bugs and documented them for programmer's correction.
- Reported customer feedback to programmers on program weaknesses and strengths.

1989–90 **Asst. to Dir. of Production** HOMES & LAND PUBLICATIONS , Miami, FL

1987–89 **Mail Clerk** STATE OF FLORIDA, Tallahassee, FL

LYDIA SILVERS
912 Blue Mountain Road
San Ramon, CA 94583
(510) 783-6221

Objective: Position as a developmental or applications engineer, for new plastics products or packaging.

SUMMARY

- Five years' professional experience, coupled with strong education and training in plastics engineering.
- Goal oriented, creative and resourceful.
- Proven effectiveness in coordinating and teamwork.
- Communicate equally well with technical and business staff.

PROFESSIONAL EXPERIENCE

1992–present **Packaging Engineer** THE CLOROX CO., Oak Hill, CA

Design/Development

- Custom designed an innovative package for a new product which was more effective and had more commercial shelf impact, at a cost savings, over proposed alternatives.
- Developed effective interim solution to packaging problem which could be applied immediately without disrupting on-going marketing.
- Designed six packaging/advertising label options that provided needed billboard effect, and which were both commercially feasible and cost-effective.

Teamwork/Communication

- Improved working communication among technical center departments by organizing a product/package compatibility Task Force, which was able to predict, solve and avoid problems by pooling their knowledge.
- Acted as technical liaison, coordinating package development with product development, process development, marketing research, advertising, analytical, regulatory, quality assurance, legal, buying, manufacturing, contract packers, and outside suppliers.

Project Administration

- Successfully organized a packaging project in record time to meet a marketing deadline:
  – Developed project time line – Researched materials – Wrote specifications
  – Located suppliers – Coordinated in-house departments and outside vendors.
- Supervised trained and evaluated:
  ...summer intern ...technical assistant ...temporary technicians.
- Qualified materials, suppliers and contract packers; and wrote specifications.

– Continued –

Lydia chose a unique format combining features of chronological and functional style resumes. Her most recent job is described in terms of the major skills relevant to her job target, while the less relevant prior work history is played down. Full-time parenting and student status are spelled out here, to present a no-gaps work history.

PRIOR WORK HISTORY

1987–92	Graduate Student	University of Lowell – Lowell, MA
1987–89	**Teaching Assistant**	PLASTICS MATERIAL TESTING LAB, Univ. of Lowell
1985–87	Undergraduate	University of Lowell – Lowell, MA
1982–85	Full-time Parent	
1981	**Building Manager**	CAMELOT COURT apartment complex, Lowell, MA
1978–81	Undergraduate	Lowell Technological Institute, Lowell, MA

EDUCATION & AFFILIATION

M.S., Plastics Engineering – University of Lowell, 1992
B.S., Plastics Engineering – University of Lowell, 1987
Member, Society of Plastics Engineers
Graduate courses:

- Polymer Structure, Properties and Applications
- Product Design
- Plastics Processing Theory
- Adhesives and Adhesion
- Coatings

Rochelle's intent is very clear.

ROCHELLE NORMANDY

(408) 577-3665

1256 Richlieu Drive
San Jose, CA 95129

Specialist in PC support services for large and small businesses

PROFILE

- 12 years' experience in the computer field; directed PC training and support for 11 National Computer stores.
- Infectious enthusiasm for computers; gifted and inspiring PC trainer.
- Expert trouble-shooter and problem solver.
- Proven ability to design systems for quick access to vital information.
- Committed to excellent service and customer satisfaction.

PROFESSIONAL EXPERIENCE

Training

- Trained thousands of individuals at all levels, in computer literacy: corporate executives, accountants, sales people, clerks.
- Developed and presented trainings for clients throughout the western U.S., e.g.: manufacturing, service agencies, banking, telecommunications, and government.
- Managed the design and implementation of customized hands-on classroom trainings and computer labs:
  - Met with department heads to clarify specific training and scheduling needs;
  - Provided software and hardware for hands-on training in: computer fundamentals, word processing, spreadsheets, relational databases, communications, graphics, macro programming; labs in customizing spreadsheets and databases.

Systems Design and Implementation

- Designed computerized workstations for cost-effective use of existing hardware, scheduling for multiple-user access and coordinating related functions.
- Designed and computerized an inventory system for 6 National Computer stores, increasing inventory turnover by 60%.

Maintenance Contracts

- Provided field service maintenance on a 24-hour basis, including telephone hot-line support and routine testing and maintenance for all major computer products: ...IBM ...Compaq ...Apple ...Epson ...AT&T ...Okidata.
- Successfully identified service problems in initial service call 90% of the time.

EMPLOYMENT HISTORY

1990–present	*Director of Training Operations*	NATIONAL COMPUTERS, San Francisco
" "	*Sales Manager*	" " "
" "	*Customer Support Manager*	NATIONAL COMPUTERS, Silicon Valley
" "	*Service Manager*	NATIONAL COMPUTERS, Monterey
1984–89	*Senior R&D Technician*	MONARCH SYSTEMS, San Jose
" "	*Production Supervisor*	" " "

EDUCATION & TRAINING

Business Administration – San Jose State University

Dale Carnegie Sales and Management courses

• IBM Service School • Apple Service School • Compaq Service School

CORONET GIBSON

1416 Acton Street
Berkeley, CA 94702
(510) 838-6755

OBJECTIVE

Position as Coordinator of Research Services for DataPro Corporation

SUMMARY

- 5 years' experience as data manager at UC Berkeley.
- Outstanding teacher; taught several RAs to use the UC computer.
- Background in psychology.
- Skill in refining and translating researchers' goals into computer languages.
- Dependable and conscientious; accurate at detail work.
- Easy to work with; a cooperative and supportive colleague.

RELEVANT SKILLS & EXPERIENCE

DATA MANAGEMENT

- Created and maintained highly accurate social science database for UC Psychology Department's Stress & Coping Project, directed by Richard Lazarus:
  - Assisted researchers in phrasing questions to assure useable data;
  - Supervised interviewers' field work;
  - Designed coding manuals and supervised coding and data entry;
  - Cleaned data; prepared code books for documentation and ready access to data.

TEACHING/TRAINING

- Successfully taught computer logic to Research Assistants who had no previous experience.
- Trained psychology students in editing and file management.
- Taught RAs the detailed use of technical software, including:
  - How to translate research goals into concrete commands;
  - How to de-bug their own programs;
  - How to effectively use the programming manual.
- Assisted faculty in establishing research goals consistent with the data available.

USE OF UC COMPUTER SERVICES

- Effectively consulted with UC Computer Center's technical consultants to resolve a wide range of problems.
- Opened and renewed computer accounts on the UC system, including faculty approval and allocation of disk space.

RELATED SKILLS

- Working knowledge of UNIX system, and data management software packages.
- Used Lotus 1-2-3 as consultant for UC faculty.

EMPLOYMENT HISTORY

1990–present	*Research Associate*	PSYCHOLOGY DEPT., UCB, Stress & Coping Project
1982–90	*Research Assistant*	SOCIAL ACTION RESEARCH CENTER, San Rafael
	" "	CALIFORNIA CONNECTION, Berkeley
	" "	INST. FOR RESEARCH IN SOCIAL BEHAVIOR, Berkeley
1977–84	*Research Assistant*	Prof. P.W. Sperlich, UCB Political Science Dept.

EDUCATION

B.A., Psychology, UC BERKELEY; Berkeley, CA
Graduate studies: Clinical Psychology, UC BERKELEY
Advanced Data Management, UC Extension

NORMAN WHITMORE
1717 Rutgers Avenue
Oakland, CA 94602
(510) 657-3201

When applying for a non-university position, Norman could drop the Ph.D. from his resume so as not to appear over-qualified.

Objective: Scientific/technical writing position with a corporation or university.

PROFILE

- Published writer; 7 years' experience in university teaching.
- Highly effective in communicating with engineers and technicians, and translating scientific information into everyday language.
- Drafted user and features guides for technical products and systems.
- Analyzed corporate telecommunications needs and wrote customized proposals.
- Successfully managed and developed major telecommunications accounts.

PROFESSIONAL EXPERIENCE

Writing

- Authored book, articles and reviews; edited textbook series.
- Produced reports and business plans for management, and proposals for clients.
- Wrote user guides and feature summary sheets for electronic equipment.

Technical Analysis and Presentation

- Analyzed the capabilities and features of new electronic equipment and products for marketing to distributors and the buying public.
- Identified and organized the telecommunications needs of senior management and their engineering, legal, financial and marketing staff. Followed up with oral and written presentation of customized proposals.

Project Management

- Independently developed successful marketing plans for telecommunications manufacturer, which quadrupled sales in the western U.S.
- Reported to corporate management, summarizing opportunities and problems with new products, and recommending marketing and technical improvements.
- Supervised and trained sales representatives.
- Trained client staffs and sales teams in the use of new technical equipment.
- Discovered and developed profitable new market of previously unaffiliated independent phone companies, through extensive field research

EMPLOYMENT HISTORY

1993–present	***Manager Distribution Sales***	BRUENER COMMUNICATIONS, San Mateo, CA
1992–93	***National Accounts Executive***	BEAM CORP, Digital Telephone Systems Division, San Mateo, CA
1991–92	***Major Accounts Executive***	ULTRA DYNAMICS COMMUNICATIONS CO., Foster City, CA
1989–91	***Senior Sales Representative***	XEROX CORPORATION, San Carlos, CA
1982–89	***Lecturer in History***	UNIVERSITY OF SYDNEY, Sydney, Australia

EDUCATION

Ph.D., German History – University of Wisconsin
M.A., Economic History – University of California, Berkeley
B.A., Economic History – Rutgers University

TECHNICAL / COMPUTERS

Mark presents a smooth blending of work history and education in Iran and the US.

MARK EBRAHIMI

89 - 12th Street, Apt. 144
Portland, OR 97207

(503) 234-5678

Objective: Position as telecommunications customer service/supervisor.

SUMMARY

- Promoted to increasing level of responsibility within Telecommunications Dept.
- Extensive knowledge of repair problems for both old and new telephone systems.
- Sincerely enjoy the challenge of providing high quality direct service to clients.
- Can be trusted to handle end-user complaints diplomatically and efficiently.
- Excellent working relations with co-workers.

RELEVANT EXPERIENCE

Customer Service

- Received and successfully handled thousands of repair calls from university departments:
  - Screened calls and made preliminary technical assessment of the problems;
  - Referred calls to appropriate vendors: ...Pac Bell for line problems ...AT&T for set problems ...new vendors for new electronic sets;
  - Followed through aggressively with vendors to assure that good service was delivered.
- Reduced university time and expense on repair orders by:
  - Minimizing time spent on each incoming call;
  - Providing increased level of technical advice to telephone users on appropriate use of features of their new electronic sets.

Technical Skills

- Researched technical information using:
  - Computer terminal database to identify line features;
  - Technical manuals to assist in diagnosing line problems.
- Implemented new comprehensive record keeping system, documenting the repair history of each phone line, as well as their disposition codes, to increase the quality of service.
- Inspected pagers, making preliminary diagnosis, and referred repairs to vendors.

EMPLOYMENT HISTORY

1995–present	***Telecommunications Rep***	Univ. of Oregon, Telecommunications Dept.
1994–95	***Tech. Clerk/Telecommunications***	Univ. of Oregon, Telecommunications Dept.
1993	***Teacher – English as a 2nd Lang.***	Iran / America Society – Tehran, Iran
1985–93	***Translator/Interpreter***	Bureau of Translators – Tehran, Iran
1983–85	***Telemedia Supervisor/Teacher***	Bell Helicopter – Tehran, Iran
1981–83	***Teacher – English as a 2nd Lang.***	Air Force Language Academy – Tehran, Iran

EDUCATION & CREDENTIALS

B.A., Psychology, 1988 – Teacher Training University; Tehran, Iran
California Community College Teaching Credential, 1994
Completed computer programming course, Univ. of Oregon Extension
Tri-lingual in English, Spanish, Persian

See John's cover letter on page 270.

JOHN BRIDGES

97 Foothill Lane • Berkeley, CA 94705 • (510) 990-3466

Objective: Research Associate position with a biotechnology firm / basic research lab, focusing on immunology and product development.

SUMMARY

- Highly inquisitive, creative and resourceful.
- Excellent skills in communication and collaboration.
- Skilled in all phases of hybridoma production.
- Good working knowledge of immunology.
- Inspired by the challenge of research and experimentation.

PROFESSIONAL ACCOMPLISHMENTS

Applied Research

- Successfully developed new antibodies for use in breast cancer research and therapy:
  - Experimented with antigen preparation and immunization routines, resulting in the desired immune response;
  - Carefully monitored the antisera to ensure presence of desired B-cell population;
  - Tailored screening strategies using ELISA, RIA and Immunoblot techniques, to effectively isolate the desired hybridomas.
- Developed, in collaboration with others, a novel assay which identified the antibodies' ability to bind to live, intact tumor cells.
- Delivered periodic presentations of results and works in progress, to staff of Cancer Research Institute.

Innovation/Exploration

- Pursued unique opportunities for experimentation, for example:
  - Researched and worked out procedures for in vitro immunization of human lymphocytes;
  - Explored and experimented with hybridoma production, using lymph node cells of a cancer patient;
  - Experimented to induce animals' immune system to respond to a weak antigen.

Lab Skills

- DS-PAGE Electrophoresis
- Radiolabeling of Antibodies
- Tissue Culture
- Affinity Chromatography
- Immunoblot Strip Assay
- Hamster Egg Penetration Test
- Electroblotting
- Lyophilization
- Isotyping

EMPLOYMENT HISTORY

1994–present	**Lab Technician**	SCHILLING CANCER RESEARCH INST., Berkeley
1990–94	Full-time student	UC Santa Barbara
1993 summer	**Research Asst.**	UC Santa Barbara Biology Dept.
1989	**Youth Counselor**	RAINBOW RIVER DAY CARE PROGRAM, LA
1988	**Emergency Med. Tech.**	SEALS AMBULANCE, Costa Mesa
1987	**Teaching Assistant**	ALTA VISTA ELEM. SCHOOL, Redondo Beach

EDUCATION

B.A., Cell Biology & Physiology – UC SANTA BARBARA, 1994
Related coursework: Immunology & Lab, Biochemistry, Virology, Microbiology

LARRY B. ELTON
2330 Blake Street
Berkeley, CA 94704
(510)437-4288

Objective: Senior position in engineering management.

SUMMARY

- ✔ Business oriented; able to understand and execute broad corporate policy.
- ✔ Strength in analyzing and improving engineering and administrative methods.
- ✔ Effective in maintaining productivity and management/team communication.
- ✔ 17 year track record in managing both large and small workforces.
- ✔ Successful in negotiating favorable design and construction contracts.

EXPERIENCE & ACCOMPLISHMENTS

1989–present ***Project Manager*** ATLANTIC RICHFIELD CO., Walnut Creek, CA

- Supervised recruitment and staffing of over 40 project team professionals.
- Wrote detailed execution plans for major design and construction projects, involving:
  – project staffing – preliminary schedule – preliminary cost estimate
  – engineering drawings – construction contractor selection – definitive cost estimate
  – approvals of contractor construction plans.
- Wrote 800-page Construction Management Guide documenting standardized construction procedures and reporting.
- Increased productivity 12% by introducing a popular 4-day/48-hour work week alternative.

1987–89 ***Project Manager*** ALLIED CHEMICAL CO. (now Allied Signal), Morristown, NJ

- Developed innovative, cost-effective concept in project management of specialty chemical plant, assigning the design engineering to outside contractors.
- Wrote comprehensive summary for senior level management, incorporating monthly reports from specialty engineering, project engineering, and construction management.

1984–86 ***Project Engineer*** SUN OIL CO. (now Sun Co.), Philadelphia, PA

- Successfully headed off loss of over a million dollars, due to potential business failure of primary contractor, by negotiating directly with subcontractors.
- Trained 25 skilled salespeople to effectively demonstrate patented equipment to various industries.

1977–84 ***Engineer*** PENNWALT CORP., Philadelphia, PA

- Conceived and patented highly profitable design for a Refrigerant Recovery System which realized a profit of over $15 million in a period of 5 years.

1974–77 ***Captain*** U.S. Army Infantry

EDUCATION

B.Sc. Ch.E., Chemical Engineering – UNIVERSITY OF WASHINGTON, Seattle, WA
Graduate studies – PENNSYLVANIA STATE UNIVERSITY and TEMPLE UNIVERSITY

Patricia Delich

Instructional Designer/Trainer specializing in online technologies

P.O. Box 3289, Berkeley, CA 94703 ▲ (415) 359-5804
pdelich@ix.netcom.com ▲ http://www.gene.com/ae

QUALIFICATIONS

▲ Train-the-trainer expert in building educational virtual communities on the Internet.

▲ Authored print-based and electronic instructional materials in support of educational communities.

▲ Skill in translating complex concepts into learning-friendly educational materials.

▲ Reputation as the technology "Dear Abby" for a national online community.

EDUCATION

Certificate in Telecommunications, Cal State Hayward University, 1995

MA, Instructional Technologies (Honors), San Francisco State University, 1994

BA, Psychology (Honors), John F. Kennedy University, Orinda, 1992

RELEVANT ACCOMPLISHMENTS

▲ Designed, developed, and delivered online Internet training seminars nationwide for: National Science Teacher Association, National Association of Biology Teachers, Biotech on a Shoestring, Teach for America, Science Education Network Academy, Genentech and others.
- Training sessions covered topics such as: how to get connected to the Internet using dial-up access; using Netscape; and ways to use the Internet in the classroom.
- Trained national teacher leaders to co-facilitate the training sessions.
- Devised creative ways to work around connection problems in a variety of training situations: hotels, universities, classroom settings and convention halls.
- Spearheaded a long term cost saving proposal for connecting 23 computers for all day convention hall training sessions.

▲ Created electronic and paper-based instructional materials and publications. Portfolio available.
- Experienced with instructional design cycles from needs assessment to evaluation.
- Worked closely with end users to customize training materials.
- Wrote several publications on how to connect and use Internet technologies.

– Continued on page two –

RELEVANT ACCOMPLISHMENTS (continued)

▲ As **Manager of Online Technologies** for Access Excellence—a national online science education program sponsored by Genentech:

- Managed an online community on America Online and the World Wide Web.
- Purchased and configured over a million dollars of computer equipment.
- Liaison between Genentech and America Online, Apple Computer, and Netcom.
- Formulated and implemented a plan to move 300 users from AOL to an ISP.

▲ Pioneered the **Video Conference Coordinator** position for Lawrence Berkeley Laboratory:

- Coordinated worldwide communications between LBL and multiple remote sites.
- Trained and assisted local and remote users.
- Part of a team that formulated policies used in the worldwide user community.
- Oversaw the daily operation and coordination of the technical staff.

▲ **Rich consulting experience** from interactive television to teaching (Cal State Hayward University and Academy of Art College) to online development for Apple Computer, Inc.

EMPLOYMENT HISTORY

1995–present	Manager, Online Technologies	**Genentech, Inc.**
1994–1995	Video Conference Coordinator	**Lawrence Berkeley Laboratory**
1989–1995	**Computer Consultant**	San Francisco Bay Area

AWARDS

Genentech	**Distinguished Contribution**, 1995, 1996 – Significant contribution to project/process. – Extraordinary productivity/workload.
SFSU	**Distinguished Student**, 1993–94 – Distinguished achievement in Instructional Technologies.
SFSU	**Academic Excellence**, 1993

MICHAEL HAYAKAWA
345 Thornhill Drive
San Francisco, CA 94141
(415) 212-4444

Objective: Position as Director of Imaging Services.

PROFILE

- Special talent for establishing rapport with both management and support staff.
- Highly effective in developing a positive and productive work environment.
- Ability to get the job done, hold expenses down, and increase profitability.

WORK EXPERIENCE

1991–present **Radiology Dept. Manager** SPAULDING MEMORIAL HOSPITAL, Oakland

Supervision and Employee Relations

- Upgraded staff skill level and productivity, and minimized staff turnover:
  - Designed more effective in-service and cross-training programs;
  - Demonstrated a strong personal interest in individuals' skill development.
- Increased staff morale and built team spirit through strong support of productive staff members.

Marketing and Profitability

- Steadily increased contribution margin of radiology department.
- Increased outpatient referral rate 15% by providing same-day diagnostic report.
- Introduced mammography screening to the spectrum of department services; developed a marketing/pricing strategy resulting in immediate increase in referrals.

Planning and Development

- Installed new Phillips LX CT Scanner:
  - Evaluated all options and selected the Phillips scanner for its diagnostic capabilities;
  - Oversaw installation, working with architects and construction crews;
  - Coordinated and arranged for interim CT service during installation;
  - Designed and implemented in-service training on new scanner.
- Advised on the installation of a new radiology department computer system, and the subsequent in-service training for staff.

1989–90	Full-time student	UCLA
1984–88*	**Evening Supervisor**	LOS GATOS-SARATOGA COMMUNITY HOSPITAL
1984–88*	**Evening Supervisor**	SANTA TERESA COMMUNITY HOSP., San Jose
1983	**Staff Technologist**	BROADWAY HOSPITAL, Vallejo
1979–82	**Asst. Chief Technologist**	DAVID GRANT MEDICAL CENTER, Travis AFB

*concurrently

EDUCATION

M.B.A., Management – UNIVERSITY OF CALIFORNIA, BERKELEY
B.S., Business Administration – UNIVERSITY OF CALIFORNIA, LOS ANGELES

PROFESSIONAL AFFILIATIONS

- American Hospital Radiology Administrators
- CA Society of Radiologic Technologists
- Certified Radiologic Technologist
- American Registry of Radiologic Technologists

Education Resumes

Nicholas Stavrinides	Educational program development	110
Kathleen Webster	Library assistant	112
Tudy Larkspur	Career counseling	113
Charles LaBuz	School counselor, secondary	114
Christiane Lloyd	Instructor, language	116
Fereshteh Ashkani	Childcare or teaching, daycare	117
Sharon Zimmerman	Art teacher	118
Elizabeth Woolsey	Educational program development/computer	120
Mariana Kadish	Translator/interpreter, Spanish	122
Martha Jupiter	Teacher/children's mental health	124
Michelle Olson	Workshop presenter, career counseling	125
Sarah Whitaker	Adult education teacher	126
Sandra Dietz	Substitute teacher	128

NICHOLAS A. STAVRINIDES

3331 Hermit Way
Santa Rosa, CA 95405
(707) 571-8928

EDUCATION CONSULTANT

Specializing in Educational Program Development in the Arts, Media and Travel

PROFESSIONAL PROFILE

- 15 years' experience promoting the role of art and history in liberal arts curricula through educational program development, teaching, traveling, and research.
- Demonstrated expertise, academic credentials, specialized training, and research skills sought after by companies developing educational media, film and documentary projects.
- Excellent communication and writing skills necessary for articulating and formulating ideas useful to the development of creative educational multimedia, programming and presentation.
- Passion for teaching and commitment to education enriched by overseas studies and years of travel.

PROFESSIONAL EXPERIENCE

Educational Program and Curricula Development

- Five years' experience developing, organizing, and conducting educational tour programs to Europe and the Middle East, focusing on the invaluable role of art and history as cultural enrichments to our lives.
- Developed and introduced an Interactive Teaching Program in the study of art, geography and history for middle school students.
- Launched a committee effort to improve English Composition Board standards for undergraduate liberal arts curricula.
- Co-wrote and helped produce a series of art videos for use as educational tools.
- Designed curricula and taught university courses in Classical Mythology, Greek and Roman Art and Archaeology, Ancient Athletics, Latin and Biblical Greek.

Lecturing and Writing

- Delivered lectures to educators, students, professional groups and church organizations on archaeology, art, history and religion.
- Travel guide and lecturer to students and professional groups on over 25 educational tours to Europe and the Middle East.
- Wrote a series of travel commentaries, handouts and walking-tour booklets for travelers exploring Western art and culture.
- Wrote catalogue entries and descriptions for special museum exhibits.
- Wrote career profiles for business professionals, and edited employee manuals.

– Continued –

Fine Arts Media Presentation, Design and Field Research

- Created and delivered multimedia-based lecture series on history and culture, highlighted by presentations incorporating footage and still-photography, drawn from motion picture epics, television documentaries and personal collection of over 8000 travelogue slides.
- Collaborated on the design of educational interactive museum displays.
- Four years archaeological field work and research at ancient sites in Greece and Israel, including analysis, mapping, and restoration of historic monuments; conservation and cataloging of artifacts.
- Organized and held two private art exhibitions to promote and market internationally recognized Polish and Czech artists.
- Production assistant on live-action short film featured at Houston Film Festival.

EMPLOYMENT HISTORY

1993–present	**International Tour Director**	GRAND CIRCLE TRAVEL, Boston, MA
1989–present	**Art Consultant**	ARTTREK INT'L, San Francisco
1989–1992	**Art Consultant**	CALDWELL SNYDER GALLERY, San Francisco, CA
1986–1988	**Financial Officer**	DMB SECURITIES, San Rafael, CA
1984–1986	**Credit Analyst**	WELLS FARGO BANK, San Francisco, CA
1979–1983	**Instructor**	UNIVERSITY OF MICHIGAN, Ann Arbor, MI

EDUCATION AND TRAINING

Ph.D. candidate, M.A., Classical Art and Archaeology, 1984/1982
UNIVERSITY OF MICHIGAN, Ann Arbor

A. B., Classical Languages, *summa cum laude*, 1979
UNIVERSITY OF CALIFORNIA, Berkeley

Undergraduate Study and Field Research Program, 1978
AMERICAN SCHOOL OF CLASSICAL STUDIES, Athens, Greece

Certified and Licensed International Tour Guide, 1993
INTERNATIONAL TOUR MANAGEMENT INSTITUTE, San Francisco

Professional Affiliations: AIA, APA (American Philological Association)

Languages Skills: Modern Greek, Italian, German, French, Latin, Ancient Greek

Computer Literacy: Experienced in operating Apple IIe, Performa 63OCD and IBM PCs.

– Resume written by Nicholas Stavrinides –

KATHLEEN WEBSTER
8022 Pine Street
Oakland, CA 94606
(510) 909-4334

Objective: Position as a Library Assistant.

SUMMARY

- Over six years' experience working in library settings.
- Smart and personable; interact easily with the public.
- Strong and precise clerical skills.
- Capable in research and reference work.

EMPLOYMENT HISTORY

1989–present **Library Aide** PUBLIC LIBRARY – Downtown branch, Hayward, CA

- Interacted with the public in all aspects of library work:
  – Processed applications and worked in the registration file;
  – Retrieved reserve books; shelved books; organized book shelves;
  – Typed book orders; processed new books; repaired damaged books.

1996 **Instructional Asst.** MILLS COLLEGE, Oakland, CA

- Operated Apple Macintosh computer and Laserwriter printer, as Instructional Assistant for Mills College graduate course in Technical Communications.

1992–present **Store Merchandiser** GIBSON CARD CO., Oakland, CA

- Inventoried, ordered and stocked merchandise for national greeting card company, interacting with personnel at three store locations.

1989 **Census Worker** U.S. CENSUS BUREAU, Oakland, CA

- Tabulated regional statistics; organized and dispatched census forms.

1988 **Library Assistant** FEATHER RIVER COLLEGE, Quincy, CA

- Answered simple reference questions of library patrons.

1987 **Fire Control Clerk** U.S. DEPT. OF FORESTRY, Quincy, CA

- Performed clerical duties and acted as Relief at the Fire Control Dispatch Center.
- Dispatched fire fighting equipment from a warehouse distribution center.

EDUCATION

B.A., English (with honors) – Mills College, Oakland, CA, 1996
A.A., Social Sciences (with honors) – College of Alameda, Alameda, CA, 1994

Tudy generalizes about her prior years of work, having already detailed the most recent 10-year period.

TUDY LARKSPUR

1219 Snowcrest Road • Pleasant Hill, CA 94523 • (510) 141-6559

Objective: **Career counseling position** with St. Elizabeth's College career center.

Summary

- Graduate degree in Career Development from JFK University.
- Sensitivity and empathy for people in career exploration.
- Experience in counseling both students and graduates.
- Enthusiastic and committed to professional excellence.

RELEVANT EXPERIENCE

Counseling/Interviewing

- Counseled individuals at JFK University Career Center for nine months, covering:
  ...test interpretations ...job search skills ...resume preparation
  ...interviewing techniques ...networking skills ...researching employers.
- Administered testing and evaluation instruments, such as Strong-Campbell Interest Inventory, Values Card Sort, Myers Briggs Type Indicator.
- Counseled groups and individuals at University Y's Turning Point Career Center:
  - Assessed needs of drop-in participants, providing informal counseling and guiding them to appropriate resources;
  - Led Job Search Support Group and workshop on Goals and Decision Making.

Resource Development

- Conducted in-depth research of a corporation to determine its desirability as a potential employer:
  - Read annual reports and visited company headquarters to observe working conditions, learn history of organization, observe relationship to community and determine potential as a career path.
- Located and directed clients to career library resources, such as professional publications, labor market information, college catalogs, Bay Area job outlook reports, and information on special populations (handicapped, refugees, etc.).
- Attended Employer Information Series at Turning Point, increasing specific knowledge of entry requirements and working conditions in various career fields.

WORK HISTORY

1995–present	**Counseling Extern**	Univ. YWCA/Turning Point Career Ctr., Berkeley
1993–96	Student	John F. Kennedy University, Orinda, CA
1991–92	**Part-time paralegal**	Moraga law firm
1989–90	**Paralegal intern**	Walsh, Morton, Meaden law firm; Moraga, CA
1989–90	Student	St. Mary's College, paralegal studies
1986–88	**Receptionist**	Women's Career Center, Diablo Valley College, Pleasant Hill

Prior years: variety of positions as medical assistant, receptionist, secretary; followed by 20 years of family management.

EDUCATION

M.A. Career Development – John F. Kennedy University, Orinda
Paralegal Certificate – St. Mary's College, Moraga
B.A. Liberal Arts – UC Berkeley

Chronological/functional combination

CHARLES D. LaBUZ

7420 East Mule Circle
Prescott Valley, Arizona 86314
(520) 775-2275

Objective: Position as a Secondary School Counselor

Summary of Qualifications

- Seventeen years' experience in counseling and teaching.
- Sincere commitment to the welfare of the student.
- Special talent for assessing individual needs.
- Work supportively with colleagues and administration.

GUIDANCE & COUNSELING EXPERIENCE

1990–present *Middle School Counselor* — Granite Mountain Middle School Prescott, AZ
1988–90 *Guidance Counselor, K-8* — Middle/Elementary School, Camp Verde, AZ
1982–87 *Junior High School Counselor* — Windsor NY Junior/Senior High School

–Individual Counseling

- Counseled students on a wide range of issues: -physical, sexual and emotional abuse/neglect; stress due to divorce, separation and death; drug and alcohol abuse; suicide intervention; runaway intervention; dropout prevention; low self-esteem.

–Group Counseling

- Issues included: -divorce, separation and blended families, self-esteem, anger management, career exploration/planning, dropout prevention, peer-death bereavement.

–School Counseling

- Advised special-need students, e.g., learning disabled, gifted students, athletes.
- Developed IEPs, coordinating intervention with Committee on Special Education.
- Designed and coordinated Orientation Program for incoming middle school students.
- Competent in using and revising Master Schedule.

–Career Development

- Created and taught "Life and Career Skills" course for Junior High students, focusing on decision making skills, values clarification, and communication skills.
- Counseled at-risk students in career options.
- Introduced students to the use of computer programs for college search, and library resources on career decision making.

–Measurement & Evaluation

- Administered and interpreted a full range of instruments, including:
  -Iowa Test of Basic Skills -Stanford Achievement Test -California Achievement Test
  -Orleans-Hanna Algebra Prognosis Test -Otis-Lennon Mental Abilities Test
  -Differential Aptitude Test -Career Decision Making System
  -Strong-Campbell Interest Inventory -Gessell School Readiness Screening Test.

–Coordination of Resources

- Served as Special Education Liaison, Prescott Child Study Services.
- Collaborated with Northern Arizona University Special Grants Staff to implement Career Day for grades 5-8, and conducted parent workshops on vocational education.
- Worked closely with Northern Arizona University NCAC staff to develop and initiate group support services for Native American students grades 6-9.

– Continued –

–Coordination of Resources (continued)

- Initiated consultations with parents, faculty, and appropriate professionals.
- Referred students to community resources when necessary.
- Coordinated the intervention services of the Department of Social Services, Office of Mental Health, community agencies, and health care professionals.
- Collaborated with teachers and principals in cluster/middle-school program teamwork.

1982 *Advising Assistant/Grad Student* UNIVERSITY OF MONTANA
- Counseled undergraduate students on academic and personal problems.

1981 *Counselor Intern* BIG SKY HIGH SCHOOL - Missoula, Montana

TEACHING

1981–82 *Substitute Teacher* MISSOULA COUNTY HIGH SCHOOLS - Missoula, MT
1979–80 *Elementary Science Teacher, Grades 4-6* DARBY SCHOOLS - Darby, MT
1975–79 *Elementary Science Teacher, Grades 1-6* DARBY SCHOOLS - Darby, MT
- Coordinated science program for elementary grades.

COACHING

1973–present *Coach* for a variety of sports including: -wrestling -football -track
1993–95 Coordinator/Head Coach – Special Olympics, track and field

MILITARY SERVICE

1963–67 *Personnel Specialist* – UNITED STATES AIR FORCE

EDUCATION & TRAINING

M.Ed., Guidance and Counseling - UNIVERSITY OF MONTANA 1982
B.S., Elementary Education - STATE UNIVERSITY OF NEW YORK, ONEONTA 1973
Minor in Physical Education - NORTHERN MICHIGAN UNIVERSITY 1975
Completed additional graduate hours for permanent New York certification 1986

Recent professional seminars and workshops:

1995 Critical Incident Stress De-briefing
1994 Competency Based Guidance Program Training
1994 Strengthening At-Risk Students' Achievement and Behavior
1993 National Student Assistance Program and Group Facilitator Training
1992 Fifth Annual National At-Risk Conference
1991 Arizona Vocational Education Conference

PROFESSIONAL AFFILIATIONS

Arizona School Counselors' Association
New York State Association for Counseling & Development, New York
Broome Tioga Counselors' Association, New York

CERTIFICATIONS

Arizona	School Counselor (K-12)
Montana	Guidance & Counseling (K-12)
	Elementary Education (K-9)
	Physical Education & Health (K-12)
New York State	School Counselor (K-12)
	Elementary Education (N-6)
Washington State	School Counselor (K-12)

Here's a layout trick that makes an elaborate Job Objective look simple and clear. Note in the Education, "equivalent to masters."

CHRISTIANE LLOYD

980 - 43rd Avenue
San Francisco, CA 94121
(415) 459-6906

OBJECTIVE

Position as language instructor in German, preferably including:
- academic counseling;
- design of teaching materials;
- curricula development;
- course evaluation.

SUMMARY

- 5 years' experience teaching German at all levels to many different target groups.
- Certified trainer for student teachers.
- Strong practical and theoretical background in developing and selecting appropriate teaching materials.
- Successful and self-confident in classroom presentation and team teaching.
- Proven effectiveness in program design and administration.

TEACHING & EXPERIENCE

CLASSROOM TEACHING

- Taught German as a Second Language in a variety of settings:
  - Beginning, intermediate and advanced students;
  - Male offenders in a correctional institution;
  - Female Spanish-speaking residents of Germany;
  - Foreign laborers in employment advancement courses.

COUNSELING/TRAINING

- Trained student teachers in the classroom, and conducted seminars focusing on didactic issues.
- Advised adult immigrant students on complex personal and academic issues:
  - immigration and employment regulations;
  - housing & landlord concerns;
  - health and medical resources;
  - entrance exams and class level placement.

CURRICULA DEVELOPMENT & COURSE EVALUATION

- Improved existing curriculum in German as a Second Language, incorporating more diversity to respond both to needs and interests of students and to knowledge gained from academic research (focused on rules of grammar and on speaking/reading/writing/listening comprehension).

WORK HISTORY

1992–96	**Teacher/Language Instr.**	HAMBURG VOLKSHOCHSCHULE, German as a Second Language Dept., Hamburg, Germany
1990–91	**Nurse Substitute/Driver**	ANSCHARHOHE EPPENDORF NURSING HOME, Hamburg, Germany
1983–90	**Full-time student** **Substitute Nurse, part-time**	UNIVERSITY OF HAMBURG, Germany

EDUCATION & CREDENTIALS

German equivalent of **Masters Degree**
Credentials to teach students through 10th grade
Relevant coursework: German, Pedagogic, Politics

EDUCATION

Fereshteh points out that she has a B.A. equivalent, and also that she's eligible for US teaching credentials.

FERESHTEH (Frances) ASHKANI

89 - 12th Street, Apt. 144
Oakland, CA 94607
(510) 575-0734

Objective: Position in child care or teaching with a private school or day care center.

SUMMARY OF QUALIFICATIONS

- Successful with the challenge of teaching groups of children.
- Patient, confident, and committed in working with children.
- Teaching credentials in Iran; eligible for equivalent U.S. credentials.

RELEVANT EXPERIENCE & SKILLS

Teaching

- Taught junior-high-age children in all mathematics subjects.
- Taught math to primary-age students.
- Tutored teenagers in natural science subjects.

Child Care/Day Care

- Supervised four- and five-year-old children, teaching them basic skills in reading and drawing.
- Currently raising my own two children, ages two and four.

EMPLOYMENT HISTORY

1994–present	Family care	
1984–94	**Jr. High Mathematics Teacher**	JEAN D' ARC SCHOOL; Tehran, Iran
1984–85 (summers)	**Nursery School Teacher**	GOLESTAN NURSERY SCHOOL; Tehran, Iran
1983–85	**Primary School Math Teacher**	NASEH PUBLIC SCHOOL; Tehran, Iran

EDUCATION

B.A, Economics (equivalent); NATIONAL UNIVERSITY OF IRAN – Tehran, Iran
Graduate training in education; MINISTRY OF EDUCATION – Tehran, Iran

EDUCATION

SHARON ZIMMERMAN

788 Alameda Street
San Francisco, CA 94122
(415) 990-3132

OBJECTIVE

Position teaching Studio Art, Art History and/or Aesthetics.

SUMMARY OF QUALIFICATIONS

- Doctorate in Fine Arts; solid academic background, through both field research and studying under great artists and scholars.
- Demonstrated success in establishing art classes and programs.
- Talent for creating a stimulating, challenging learning environment.
- Broad artistic perspective resulting from extensive travel and education on three continents.
- Active artist, personally attuned to the world of art.

PROFESSIONAL EXPERIENCE

TEACHING

- Taught painting and sculpture in a variety of settings:
  - community centers in Toronto, Paris and Madagascar;
  - special education programs in Canadian public schools;
  - private art school in Italy.
- Designed and taught an Ethno-Aesthetics course on Malagasy funeral art for art history students at the Pordenone Civic Center in Pordenone, Italy.
- Taught a course in Aesthetics at the Santa Rosa Arts Center, Pordenone, Italy.
- Presented a lecture series on Judaism and Art, sponsored by St. George Church, Paris, for theologians and scholars.

CURRICULUM/PROGRAM DEVELOPMENT

- Established a Studio Art program at the American Cultural Center in Madagascar:
  - Introduced creative arts to adults who had never held a paintbrush;
  - Presented new concepts and media to students not previously exposed to the imaginative and creative processes.
- Directed Wallaceburg Community Cultural Center, the first of its kind in the area, stimulating community interest in both art theory and studio art:
  - Wrote a grant proposal successfully winning funding from Canadian government;
  - Taught painting, photography, sculpture, and drawing; lectured on art history.

– Continued –

Sharon shows that her "D.S.A.P." is equivalent to the B.F.A. degree in the US. Then in the "Summary" she points out her "broad artistic perspective resulting from extensive travel and education on 3 continents."

SHARON ZIMMERMAN
Page two

PROFESSIONAL EXPERIENCE, Continued

RESEARCH

- Led a research program for university students and scholars interested in exploring Malagasy funeral art, familiarizing students with the ethnology and anthropology of the areas and populations under study.
- Conducted a unique research project exploring the relationship of Judaism and art, featuring a fresh, new approach to this familiar subject; developed a valuable bibliography and information network on the subject.

WORK HISTORY

1995–96	**Graduate Student**	Ph.D. program, University of Paris
1994	**Lecturer in Art**	ST. GEORGE CHURCH, Paris
1992–93	**Art Program Coordinator**	JEWISH COMMUNITY CENTER, Oakland, CA
1990–91	**Art History Instructor**	SANTA ROSA GALLERY, Pordenone, Italy
"	**Lecturer**	PORDENONE CIVIC CENTER, Pordenone, Italy
1989	**Asst. Professor, Art**	UNIV. OF MADAGASCAR (on contract)
1988	**Art Program Director**	AMERICAN CULTURAL CENTER, Madagascar
1987	**Lecturer, English**	UNIVERSITY OF PARIS, Paris
1982–86	**Full-time student**	L'ÉCOLE NATIONALE, Paris

EDUCATION

Ph.D., Aesthetics & Art History (with honors), UNIVERSITY OF PARIS 1, Pantheon-Sorbonne
D.E.A., Aesthetics & Ethnology (with honors), UNIVERSITY OF PARIS 1, Pantheon-Sorbonne
M.A., Aesthetics & Ethnology (with honors), UNIVERSITY OF PARIS 1, Pantheon-Sorbonne
D.S.A.P., BFA equivalent, L'ÉCOLE NATIONALE Superieure des Beaux Arts, Paris

ELIZABETH S. WOOLSEY, Ed.D.
CONSULTANT IN EDUCATIONAL PROGRAM DEVELOPMENT

9880 Atlantic Avenue
Oakland, CA 94609
(510) 107-8776

SUMMARY OF QUALIFICATIONS

- 20 years' experience in education: teaching, research, and program design.
- Management and administrative experience.
- Written and spoken communication skills.
- Graduate degrees in philosophy and education.

PROFESSIONAL EXPERIENCE

Educational Program Development and Presentation

- Consulted in schools at all levels, on program development and evaluation.
- Developed curriculum materials for University Without Walls and taught:
  ...English as a Second Language; ...East-West comparative philosophy.
- Designed and led workshops at the University of Maryland for:
  ...inner-city teachers; ...college of education students;
  ...community college faculty; ...women re-entering job market.

Writing and Lecturing

- Edited and produced employee handbook and procedures manual for Computer Solutions.
- Currently under contract for book on computer learning.
- Developed reports on federally funded school programs for Insight Corp.
- Wrote facilitator's manuals for university workshops in education.
- Lectured, as guest of educational institutions in the U.S. and Asia.

Management and Administration

- As Executive Director for People-To-People (a nonprofit incorporated in 30 states):
  – Recruited, trained, and coordinated activities of 10,000 volunteers;
  – Managed fund-raising and a $600,000 budget;
  – Planned and led two national conferences for regional and local coordinators.
- As Program Manager for Insight Educational Corp.:
  – Planned and led educator workshops in six major U.S. cities;
  – Organized local committees to support workshops;
  – Published report on the department's work in education.
- Managed operations/supported management during startup phase of four diverse corporations.

Computer Literacy

- Experienced in the use of word processing and database software.
- Trained in simple programming and use of the Internet.
- Evaluated educational software.

– Continued –

We don't leave a gap even in this impressive work history...filling in with travel and study.

EMPLOYMENT HISTORY

1992–present	***Educational Consultant***	RICHFIELD ENTERPRISES, Petaluma, CA
	" "	EDUCATION NETWORK, INC., Santa Rosa, CA
" "	***Operations Consultant***	COMPUTER SOLUTIONS, San Jose, CA
		HEALTH RESOURCE NETWORK, Chevy Chase, MD
		GROWING OLDER, INC., Chevy Chase, MD
1990–92	***Executive Director***	PEOPLE-TO-PEOPLE PROJECT, Richmond, CA
" "	***President***	(organized volunteer visits to institutionalized people)
1988–90	Travel and study in Europe and Asia	
1986–88	***Educational Consultant***	UNIVERSITY OF MARYLAND, Baltimore, MD
" "	***Workshop Leader***	" " "
1984–86	***Program Manager***	INSIGHT EDUCATIONAL CORP., Richmond, CA
1977–84	***Teacher, Researcher***	UNIVERSITY-WITHOUT-WALLS, Berkeley, CA
" "	***Graduate Student***	" " "

EDUCATION & TRAINING

Ed.D. Education, 1984 – University of Maryland, College Park, MD
M.A. Philosophy, 1977 " "
B.A. Philosophy, 1972 (with honors) " "

National Computer Training Institute, 1995
Women's Computer Literacy Workshop, 1994
Original Computer Camp, 1993

– References on request –

EDUCATION

MARIANA KADISH
667 Vermont Avenue
San Francisco, CA 94107
(415) 342-6112

OBJECTIVE

Position as freelance translator and interpreter in Spanish
for an agency, professional office, or nonprofit organization.

PROFILE

- Sharp insight into the subtleties of both Spanish and English, with extensive background in two cultures.
- Analytic and versatile thinker, effective in developing and carrying out ideas.
- Self-motivated, creative professional; able to work independently and also coordinate with others.
- Exceptional communication and interpersonal skills.

RELEVANT WORK EXPERIENCE

1986–present **Staff Assistant** SALVADOREAN UNITED EDITORS, San Francisco
a nonprofit literary organization.

- Edited and translated literature, business correspondence, and project proposals.
- Generated positive business response and community support for fund-raising in the arts, and on behalf of cultural events.
- Initiated ideas for Public Relations; served as spokesperson in presenting programs to large audiences.

1984–86 **Sales Agent** DEBORAH VALOMA, DESIGNER, Berkeley

1981–83 **Counselor-in-training** CORYELL & CO., Rehab Counselors, Oakland

- Administrative assistant to Rehabilitation Counselor in labor court cases for non-English speaking clients.

1980 **Paralegal** W.M. STAHL, ESQ., IMMIGRATION LAW, Oakland

- Represented Spanish speaking residents in legal and medical offices, insurance companies, and in municipal and criminal court.
- Prepared petition for residence, citizenship, and political asylum on behalf of non-residents at immigration law firm.

1979 **Translator/Co-Owner** LANGUAGE SERVICES, San Clemente

- Taught English as a Second Language for Spanish speakers.
- Interpreted at international conferences: presented simultaneous technical papers for audience of professionals; facilitated dialogue among educational conference participants.

— Continued —

EDUCATION

San Francisco State University, 1987
Pacific Basin School of Textile Arts, Berkeley, 1983–85
Fiberworks Center for Textile Arts, Berkeley, 1984–85
B.A., Latin American Studies –UC Berkeley, 1981
Emphasis in Anthropology, Literature, and Political Science

PROFESSIONAL AFFILIATIONS

- **American Translators Association,** Associate Member
- **California Court Interpreters Association,** Member
- University of California Alumni Association

Even though Martha's staying in a teaching related field, her experience looks the strongest in the functional format.

MARTHA JUPITER

219 Prince Street
Berkeley, CA 94705

Home (510) 556-8907
Msg. (510) 455-7676

Objective: Teacher/counselor in children's mental health services.

SUMMARY OF QUALIFICATIONS

- Over four years' experience teaching children and preschoolers.
- B.A. Degree in psychology with emphasis in Child Development.
- Teaching background enriched by overseas positions and travel.
- Experience with children and parents from varied cultures.
- Successfully organized and managed a summer day care program.
- Communicate with children and parents with warmth and diplomacy.

RELEVANT EXPERIENCE & SKILLS

Teaching

- Team-taught children, aged one to five years, in UC Berkeley Child Care system:
  – Provided warm, supportive environment for developing emotional and social growth;
  – Worked with small groups consisting of children of similar developmental levels.
- Taught English as a Second Language, on a one-to-one basis, to native children in Greece; originated and publicized this ESL program to serve families in the city of Athens.
- Co-organized, with six families, a summer day care program to bridge the interim period not covered by UC Berkeley system; served as sole teacher, rotating the site weekly to different homes.

Planning & Supervision

- Planned, as member of staff team, a full range of activities to assist children in advancing their social and motor development:
  ...art ...cooking ...story telling ...music ...excursions ...supervised play.
- Assisted Children's Librarian at city library in organizing reading materials and supervising children in book selection.

Parent Contact & Staff Relations

- Participated in staff meetings, addressing problems including family relations, staff cooperation, community support, and problem issues with individual children.
- Met with parents at parent/staff meetings, inviting their input into all phases of program planning and generating a cooperative atmosphere.

WORK HISTORY

1993–present	***Teacher***	self-employed English teacher/tutor in Athens, Greece; U.S.
1991–92	Travel and research in Greece and northern Europe	
1988–91	***Teacher***	UC BERKELEY CHILDCARE SYSTEM
1982–88	***Secretary***	CALIFORNIA FARMER PUBLISHING CO., SF
1981–82	***Library Aide***	Children's Room, GRACE A. DOW LIBRARY, Midland, MI

EDUCATION

B.A., Psychology/emphasis in Child Development;
graduated with high honors, UC Berkeley, 1991

Chronological/functional combination.

MICHELLE OLSON

322 Beverly Place, Piedmont, CA 94611
Home: (510) 302-5867 Business: (415) 685-1990

Objective: AHP conference presenter, for workshops in career development.

SUMMARY OF QUALIFICATIONS

- Masters in Career Counseling and Development.
- Outstanding teacher, specializing in creativity in business.
- Extensive background in counseling, instructing and program development.
- Highly creative and intuitive problem solver.
- Special talent for drawing people out and clarifying their problems and needs.

PROFESSIONAL EXPERIENCE

CONSULTANT **Michelle Olson & Associates; Oakland, CA** **1990–present**

Counseling

- Counseled and motivated individuals to recognize and understand personal needs, problems, alternatives and goals, using a combination of practical problem solving skills and a transpersonal approach to counseling. Clients report that this combination of Eastern philosophy and Western psychological systems allows them to move to new levels of understanding with consistent, practical results.

Consulting

- Designed and presented seminars on conflict resolution for business and government agencies, increasing the number of employees utilizing EAP counseling services.

Instructing

- Prepared and presented lectures for the Creativity in Business course for the MBA programs at Stanford School of Business and California State University, Hayward; this course has consistently been ranked as one of the most outstanding offered by the two schools.
- Developed and implemented courses and workshops for Piedmont Adult Education classes: –Effective Listening; –Managing Anger; –Making Good Decisions; attracting new students to the Adult School and increasing my private clientele.

Facilitating

- Co-facilitated and collaborated on "Creativity in Business" seminars for:
  - Stanford School of Business Alumni Association, Hawaii;
  - Young Presidents' Organization, Portland, OR.
- Created and led ongoing "Getting Clear" group for women, focused on improving their interpersonal relationship skills.

OWNER/DIRECTOR **Aerobic Exercise Studio; Oakland, CA** **1980–90**

Business Management

- Founded, developed and managed Oakland's first aerobic exercise studio.
- Trained and supervised 8 aerobic instructors, serving more than 300 clients.

Program Development

- Designed and presented workshops on stress management; the workshops evolved into a private practice, Michelle Olson & Associates.

EDUCATION

M.A., Career Counseling & Development – John F. Kennedy University, Orinda, CA
B.A., Marymount College; Tarrytown, NY
Secondary Teaching Credential; University of California, Berkeley

SARAH WHITAKER

13247 Colby Street
Berkeley, CA 94705
(510) 809-4245

Objective: **Position as Adult Education Teacher:**

- Arts and Crafts • Painting and Drawing • Creative Expression
- Drama • Physical Fitness • Physical Skills and Sensory Motor
- Communication Outlet • Current Events • Other Subjects

- Also available to teach/counsel adults with handicaps, developmental disabilities, or psycho-social problems.

SUMMARY OF QUALIFICATIONS

- Credentialed and experienced adult education teacher.
- Talent for incorporating 10 years' background in art, drama, and therapy into an adult education curriculum.
- Creative skill in getting students involved.
- Sensitivity to the needs of adults with physical and emotional problems, through 6 years' experience as a certified body therapist.

RELEVANT EXPERIENCE

Teaching

Taught in three adult education programs sponsored by the Oakland Unified School District:

- **"Older Adult Program"** – Held classes for adults 65+, at Senior Citizens' Centers, Hospitals and Retirement Homes;
  Subjects covered: ...Painting and Drawing ...Physical Fitness ...Ceramics.
- **"Older Adults In Care Facilities Program"** – Taught adults 65+, at intermediate and long-term care nursing facilities: ...Drawing and Painting ...Arts and Crafts ...Communication Outlets ...Physical Fitness.
- **"Adults With Exceptional Needs Program"** – Taught developmentally disabled adults in a program sponsored jointly with local community-based organizations: ...Physical Skills and Sensory Motor ...Painting and Drawing.

Counseling

- Served as on-call counselor for emotionally disturbed older adults, at Wellsprings residential psychiatric halfway house in San Francisco.
- Counseled older adults, as outreach counseling intern at Home for Jewish Parents, Oakland.
- Counseled seniors on general health issues and stress reduction, as health practitioner at South Berkeley Senior Center.
- Counseled bodywork/massage therapy clients in nutrition, stress reduction and therapeutic exercise. Made referrals to other professionals, as appropriate.

– Continued –

Sarah has very little paid experience in teaching so we used a functional format to maximize the impact of what she has done in teaching, and to show how her counseling and bodywork experience applies to her teaching goal.

WORK HISTORY

1995–present	**Adult Education Teacher**	OAKLAND PUBLIC SCHOOL SYSTEM, Pleasant Valley Adult School
1995	**On-Call Counselor**	WELLSPRING psychiatric residential halfway house for adults over 65, San Francisco
1990–present	**Bodywork Practitioner**	PRIVATE PRACTICE, Oakland
1991–92	**Health Practitioner**	SO. BERKELEY SR. CITIZENS CENTER,
1991	**Salesperson**	CLASSIC SHOP, Oakland, (women's wear)
1988–93	**Actress**	MILL VALLEY CTR FOR PERFORMING ARTS, and other Bay Area theaters
1980–89	**Artist/Painter/Printer**	Freelance, concurrent with above

EDUCATION, TRAINING & CREDENTIALS

B.F.A., Fine Arts/Painting – UNIVERSITY OF OKLAHOMA, Norman, OK, 1985
1987 Masters Program in Counseling – SAN FRANCISCO STATE

Performing Arts Training

- Jean Shelton School of Acting
- Drama Studio/Berkeley
- Playwright's Festival – Mill Valley

Classes in: ...Scene Study ...Improvisation ...Movement ...Speech ...Script Analysis

Counseling Training

Counseling Intern, Home for Jewish Parents, Oakland;
(Counseling Master's Program, San Francisco State)
Counseling Orientation, Wellspring Halfway House
Art Therapy, San Francisco State

Credentials

Adult Education, Preliminary Designated Subjects Credential
Certified Clinical Masseuse

SANDRA DIETZ

222 - 51st Street • Oakland, CA 94609 • (510) 151-3423

OBJECTIVE

Position as substitute teacher: teaching mathematics, science or health education; elementary or junior high level.

PROFILE

- Two years' successful teaching experience.
- Firsthand knowledge of cultural differences; traveled and lived in Africa, Asia and Europe.
- Sensitive to racial issues; patient and caring.
- Strong leadership skills; able to direct and make decisions.
- Deeply committed to high quality education for children.

RELEVANT EXPERIENCE

Teaching

- Taught General Science to junior high school students in Swaziland:
  – Generated high level of enthusiasm among students;
  – Overcame language barrier through creative use of visual aids.
- Taught English as a Second Language to students age 7-50 in Japan.
- Tutored Japanese student in American customs and language.

Lesson Planning/Testing

- Developed lesson plans geared to the comprehension level of junior high students.
- Prepared, administered and graded special trial exams for junior high students, preparing them to qualify for entrance into high school.

Expertise in Math, Science, Health

- Studied and excelled in college level math: Algebra, Calculus, Trigonometry, Geometry.
- Completed coursework in Chemistry and Biology; college degree in Health Science.
- Worked as Lab Technician for City of San Leandro.

Cultural/Racial Exposure

- Worked cooperatively with professionals in a racially mixed agency.
- Lived and worked for two years in Africa and Asia.
- Traveled under challenging circumstances, throughout India and Nepal.

EMPLOYMENT HISTORY

1996–present	*Administrative Asst.*	AMERICAN RUBBER CO.;
	" "	ADIA AGENCY, East Bay
1995	*Teacher, ESL*	MAGNOLIA ENGLISH School, Shizouka City, Japan
1993–94	*Science Teacher*	PEACE CORPS, Swaziland
1992–93	*Attendant for Disabled*	JEANNIE LOJO, Hayward
1989–92	*Lab Assistant*	WASTE WATER TREATMENT PLANT, San Leandro

EDUCATION

B.S., Health Science – HAYWARD STATE UNIVERSITY
Language & Cultural Training – Peace Corps, Swaziland

EDUCATION

Therapy and Social Work Resumes

Hannah Jenkins	Counselor/addictions	130
Roslyn Marcus	Clinical social worker	131
Patrick Enright	Therapist/social worker	132
Gregory Ackerman	Mental health treatment intern	133
William Ernest	Psychotherapist	134
Carol Weitzell	Program development, elderly	135
Donnette Frost	Chemical dependency intern	136
Eileen Schulman	Social worker	138
Benjamin Farber	Psychotherapist	140

Hannah focuses her resume on her volunteer work experience showing this is her highest priority.

HANNAH JENKINS
1 La Brea, Apt. 605
Forest Hills, CA 94803
(123) 555-1212

Objective: Staff position in a program providing counseling services.

SUMMARY

- Experience in counseling individuals and public speaking.
- Easily inspire trust, warmth and rapport.
- Nine-year background in volunteer work, involving teaching, supporting and motivating adults.
- Coordinated staff and activities in a 20-person department.
- Over 15 years work experience related to counseling.

RELEVANT EXPERIENCE

TRAINING / COUNSELING / PROMOTION

As long-term participant (nine years) in a **volunteer** program:

- **Spoke** before large groups (500-1000 people) as main conference speaker.
- **Counseled** individuals of various ethnic and economic backgrounds, guiding and supporting them to successfully apply the principles of addiction recovery. Worked with people in jails, hospitals, halfway houses, and recovery homes.
- **Conducted** large and small **workshops** for conference participants.
- **Promoted** increased participation in addiction-recovery programs through direct one-to-one contact and personal encouragement.

ADMINISTRATIVE SUPPORT

- **Managed** and **coordinated** litigation group of approximately 20 staff people, as "right-hand assistant" to senior partner at Silver, Pauley & Mansley. Handled all materials and logistics relating to complex legal issues. **Wrote** daily / weekly reports.
- Served as **inter-agency liaison** between CA Housing & Development Agency and other city departments: Personnel, Budget, Accounting, Purchasing, Health Services.
- **Coordinated** new office set-up for a research project of the California State Bar Association. **Assessed** job applicants, assisted in **interviewing, coordinated** installation of equipment and phones, **supervised** research, **monitored** work flow, maintained bookkeeping and filing systems.

WORK HISTORY

1990–present **Counselor/Speaker** on addictions, as a volunteer, at jails, hospitals, halfway houses, recovery homes; working with groups and individuals. (concurrent with part-time employment as canvasser and legal secretary)

1988–89	**Billing Coordinator**	SILVER, PAULEY & MANSLEY, Forest Hills
1982–88	**Administrative Coor.**	SILVER, PAULEY & MANSLEY
1980–82	**Legal Secretary**	SILVER, PAULEY & MANSLEY
1978–79	**Legal Admin. Associate**	CITY OF FOREST HILLS
1977	**Legal Admin. Assistant**	CALIFORNIA STATE BAR ASSOC., Forest Hills

EDUCATION

•Boston University •California Institute of Finance
•New School of Social Research, Forest Hills •Forest Hills City College

ROSLYN MARCUS

409 Acadia Way • Oakland, CA 94602 • (510) 350-2123

OBJECTIVE

Clinical Social Worker in the Dept. of Psychiatry of a large HMO.

SUMMARY OF QUALIFICATIONS

- Licensed clinical social worker; 16 years' professional experience.
- Demonstrated success in reaching treatment goals, with a focus on short-term therapy.
- Strong practical and theoretical foundation in a number of therapeutic and intervention models.
- Excellent skills in group, couple, family and individual counseling.
- Cooperative and supportive colleague.

RELEVANT PROFESSIONAL EXPERIENCE

ASSESSMENT & DIAGNOSIS

- Conducted hundreds of psycho-social evaluations and in-depth interviews of individuals and families.
- Assessed needs of patients involving a great diversity of issues:

–crisis intervention	–adolescent acting-out
–family role disruption	–destructive relationships
–child abuse	–families divorcing; children's adjustment
–loss and grief counseling	–adjustment to chronic illness
–acceptance of aging	–"empty nest" syndrome.

TREATMENT

- Counseled wide range of individuals and families at time of significant family crisis:
  - Assessed personal strengths and coping mechanisms of patients and families;
  - Developed and implemented treatment plans in keeping with the priorities and focus of the patients' and families' needs and resources.

GROUP COUNSELING

- Led long-term psychotherapy groups for both mixed groups and women-only.
- Designed and led unique time-limited therapy group, called "New Beginnings," for widows and widowers.
- Led highly successful weekly support/therapy groups for spinal cord injury, stroke and cancer patients.
- Collaborated effectively with many different colleagues to co-lead therapy groups.

EMPLOYMENT HISTORY

1986–present	**Dir. of Medical Social Work**	WESTCOAST CENTER Health Plan, Oakland
1976–85	**Social Worker**	KAISER HOSPITAL, Vallejo
1973–76	**Group Work Supervisor**	CTR. FOR PERSONAL / SOCIAL CHANGE, Berkeley
1971–72	**Social Worker**	INDIAN HEALTH SERVICES, Arizona
	Mental Health Consultant	" " "
1970–71	**Child Protective Svcs. Worker**	COUNTY OF SAN FRANCISCO

EDUCATION, CREDENTIALS & AFFILIATIONS

M.S.W. – HUNTER COLLEGE SCHOOL OF SOCIAL WORK, NYC • LCSW No. LH 099998
B.A., English – cum laude, Phi Beta Kappa – HUNTER COLLEGE, NYC
Member, National Association of Social Workers

PATRICK D. ENRIGHT

1919 Shaeffer Ave.
Walnut Creek, CA 94596
(510) 635 9798

Objective: Position as **therapist, social worker and/or program coordinator** working with individuals and groups, using public therapy / social rehabilitation model.

SUMMARY OF QUALIFICATIONS

- 5 years' experience counseling individuals, couples and families.
- Outstanding skill in assessing clients' needs for clinical and social rehabilitative services.
- Effective in developing programs and reaching project goals.
- Authored curriculum and training manuals for health and safety.
- Deeply committed to my clients' well being.

EMPLOYMENT HISTORY

1995–present **Counseling Intern** ELMWOOD HOMES Cooperative Apts. Program, S.F.

- Coordinated with other mental health and social service agencies to develop individualized program plans for developmentally disabled / mentally disturbed clients:
  - Referred clients for job training, medical care and additional therapy;
  - Found alternative funding for client's medical and dental needs;
  - Served as an advocate to psychiatrists, psychologists and social workers, to assure highest quality services for my clients.
- Developed a unique group therapy format for developmentally disabled adults:
  - Increased their enthusiasm, involvement and sense of empowerment by incorporating opportunities for choice and recognition of their achievements.
- Counseled both voluntary and involuntary clients, using a range of therapy modes: ...psychodynamic ...social rehabilitative ...gestalt ...humanistic ...behavior modification.
- Organized a parent support group for families of clients, supporting their independent living by improving family dynamics.

1991–94 **Counselor/Instructor** COMMUNITY HOME HEALTH CARE, Seattle

- Conducted intake interviews in community mental health agencies, selecting appropriate clients for the program.

1989–90 **Teacher/Tutor** DEARBORN PARK SCHOOLS, Seattle

1989 **Counseling Intern** OPEN DOOR CLINIC (walk-in clinic), Seattle

- Conducted couple, group and family therapy sessions using MRI Systems approach which provided for clients' maximum potential for growth.

EDUCATION & TRAINING

M.A., Counseling Psychology – JFK UNIVERSITY, Orinda **1986**
Rehabilitation & Training Center, University of Oregon at Eugene
Licensed Massage Therapist, Seattle Massage School, Seattle, WA
B.A., History – MARQUETTE UNIVERSITY, Milwaukee, WI. 1973

GREGORY ACKERMAN
7 Captain Drive
Emeryville, CA 94608
(510) 511-2323

Objective: Position as Mental Health Treatment Specialist/Intern at Contra Costa County Jail.

SUMMARY

- Deeply committed to gaining experience and expertise in psychotherapy.
- Interested in the challenge of working with the jail population.
- Masters degree, and working toward MFCC License.
- One year clinical experience with a diverse population in an outpatient counseling center.

PROFESSIONAL EXPERIENCE

Interviewing & Assessment

- Interviewed and counseled men at Center for Men's Health and Education, providing education and emotional support for partners of women having abortions.
- Conducted intake interviews and performed assessments of outpatients at Pathways, a general counseling center.

Treatment Planning & Implementation

- Developed individualized therapeutic treatment plans, applying short and long-term strategic psychotherapeutic methods.
- Conducted successful psychotherapy with a suicidal, agoraphobic client: no longer suicidal, and currently enrolled in a paraprofessional program.

Reporting

- Maintained records of therapeutic activities and client progress.
- Delivered weekly verbal reports to counseling supervisors.
- Reported diagnoses and prognoses to Dept. of Rehabilitation, as needed.

EMPLOYMENT HISTORY

1996–present	**Counselor, MFCC Intern**	LIGHTHOUSE COUNSELING CTR., Walnut Creek
1994–95	**Counseling Student Trainee**	LIGHTHOUSE COUNSELING CTR., Walnut Creek
1993	**Full-time student**	JOHN F. KENNEDY UNIVERSITY, Orinda
1990–92	**Owner/Manager**	FIRE AND ICE, jewelry store, San Francisco
1983–90	**Gem Cutting/Sales**	SELF-EMPLOYED, San Francisco

EDUCATION

M.A., Clinical Holistic Health Education – John F. Kennedy University, Orinda, 1995
M.S., Pharmacology – University of California, San Francisco, 1983
Additional coursework: Psychology – San Francisco State College, 1976
B.A., Biophysics – University of California, Berkeley, 1975

The functional resume format works well in this case, where William wants to make a job change back into psychotherapy. He explains his earlier professional work and downplays his recent explorations into other fields.

WILLIAM ERNEST

Clinical/Psychiatric Social Worker

2432 Santa Maria Avenue
San Francisco, CA 94110
(415) 590-3936

Objective: Psychotherapy position with an agency, clinic or hospital.

SUMMARY OF QUALIFICATIONS

- Licensed clinical social worker; 10 years' professional experience.
- Strong foundation in a number of therapeutic and intervention models.
- Integrated, flexible, and appropriate approach.
- Outstanding skills in assessing clients' needs.
- Committed to bringing about real, practical results in people's lives.

RELATED ACCOMPLISHMENTS

Counseling & Interviewing

- Managed large caseload as a psychiatric social worker treating voluntary and involuntary outpatient clients through the California State Department of Health Continuing Care Services program:
  - Individual and group counseling with voluntary clients;
  - Casework follow up of involuntary clients involving home visits and referrals to other agencies providing support services;
  - Implemented a model independent community halfway house for selected higher-functioning clients to supervise their development of independent living skills.
- Maintained a small private practice for two years.
- Conducted individual, group and family therapy at the Indian Health Project.

Cross-Cultural Experience

- Trained native American paraprofessionals in basic psychotherapy concepts as applied to their unique cultural heritage.
- Served as liaison to Black, Asian-American, Latino, Chinese and Samoan community as research associate in NIMH cross-cultural needs assessment study of minority elderly.
- Counseled at a Latino heroin addiction treatment facility.

PROFESSIONAL EXPERIENCE

1995–present	Real Estate Agent	CLASSIC PROPERTIES, San Francisco
1993–94	Retail Management	MACY'S, Menlo Park
1992	**Counselor/Supervisor**	THE INDIAN HEALTH PROJECT, Eureka, CA
1989–91	**Psychotherapist**	Private practice, San Francisco
1986–88	**Psychiatric Social Worker**	CALIFORNIA DEPT. of HEALTH, San Diego
1984–86	**Research Associate**	SCHOOL OF SOCIAL WORK, SDSU, San Diego
1983–84	**Psych. Social Work Intern**	VA HOSPITAL, Outpatient MH Clinic, San Diego
1982–83	**Substance Abuse Counselor**	NARCOTICS PREV. & EDUC. SYSTEMS, " "

EDUCATION & CERTIFICATION

MSW, Clinical Social Work – SAN DIEGO STATE UNIVERSITY
BA, Economics/Political Science – UC BERKELEY • **LCSW** – 1988

One of two resumes for Carol.
See also page 34.

CAROL WEITZELL

1912 Kains Avenue • Berkeley, CA 94702 • (510) 235-3980

OBJECTIVE: Program Development position working with the elderly population.

SUMMARY

- MSW degree, focusing on administration and planning.
- Three years' experience as program specialist.
- Case management experience with elderly population.
- Strength in creative problem solving.
- Outstanding ability to work with community and professional groups.

PROFESSIONAL EXPERIENCE

ELDERLY SERVICES

- Case manager for the West Contra Costa County Office on Aging:
  - Assisted elderly population in using community resources; care planning arrangements; and support in dealing with issues of aging.
- In-Home Supportive Services Social Worker, Dept. of Social Services, West Contra Costa County:
  - Implemented County Homemaker/Chore program.

COMMUNITY SERVICES

- Compiled and edited a comprehensive community resource manual instructing social workers about the services available in the city of Boston, how to access them, and procedures for qualifying.
- Liaison for community/professional Advisory Board for three years; projects of the Board included:
  - Community resource evaluation and coordination;
  - Budget plan development for child welfare agency.

ADMINISTRATION & PLANNING

- Developed an in-service training program for social work staff which increased their professional expertise and theoretical background:
  - Developed a form for self-assessment by workers;
  - Wrote and managed annual training budgets;
  - Contracted with trainers to provide instruction in specific issues in social work;
  - Met regularly with supervisors to evaluate the program on an on-going basis.
- Produced a community social service needs assessment to be incorporated in the budget narrative of a Boston DSS office.

EMPLOYMENT HISTORY

1995–present	***Case Manager***	AREA OFFICE ON AGING, W. CONTRA COSTA CO., CA
1995–present	***Social Worker***	CONTRA COSTA COUNTY, CA
1993–95	***Photographer***	Self-employed
1990–93	***Program Developer***	MASSACHUSETTS DEPARTMENT OF SOCIAL SERVICES
1989–90	***Admin. Analyst***	" " "
1988–89	***Child Welfare Worker***	" " "
1987–88	***Student Intern***	" " "

EDUCATION

MSW, BOSTON UNIVERSITY – SRS fellowship in Child Welfare, 1988

DONNETTE C. FROST
6241 Park Blvd.
Oakland, CA 94602
(510) 514-2777

Objective: Internship in Chemical Dependency Treatment Unit, working with individuals, families and groups.

SUMMARY

- 13-year professional background in both counseling and nursing.
- Special strength in educating and treating families of chemically dependent clients.
- Solid understanding of addictive and compulsive disorders.
- Sensitivity and knowledge of cultural/ethnic issues in addiction.
- Effective in balancing professionalism with sincere empathy.

PROFESSIONAL EXPERIENCE

Clinical Counseling
- Conducted long-term and short-term psychotherapy for three years with: ...individual adults ...adolescents ...groups ...families ...couples.
- Provided crisis intervention, problem solving and brief therapy in two mental health clinics and city employee assistance program.
- Counseled chemically dependent individuals and families.
- Served as outreach counselor for Community Senior Centers.

Assessment
- Conducted assessment, brief therapy, referral and follow-up for employees.
- Performed investigative intake interviews and diagnostic evaluation.
- Developed diagnostic profiles and treatment plans to address patients' therapeutic needs.

Supervision and Administration
- Evaluated and advised clinical team on psychological needs and appropriateness of services for potential clients.
- Supervised and/or trained counselor trainees, interns and community organization staff.
- Conducted educational and informational presentations on Mental Health and Chemical Dependency for community organizations.
- Coordinated and supervised an alternative disciplinary action program for students experiencing academic and emotional difficulties.

Cross-Cultural Skills
- Provided counseling services to clients from a wide range of ethnic backgrounds.
- Lived and worked with people of various cultures and social strata in Europe and the United States.

– Continued –

EMPLOYMENT HISTORY

1996–present	***Counseling Intern***	EMPLOYEE ASSISTANCE PROGRAM, City of Oakland, CA
1996–present	***Research Assistant***	INST. OF BLACK FAMILY LIFE & CULTURE Oakland, CA
1995–96	***Psychotherapist Intern***	BERKELEY MENTAL HEALTH CLINIC, City of Berkeley, CA
1994–95	***Psychotherapist Intern***	J.F. KENNEDY COMMUNITY COUNSELING CENTER, Concord, CA
1992–94	***Sr. Head Res. Counselor***	ST. GEORGE HOMES INC., Berkeley
1989–92	***Special Tutor***	SAN DIEGO UNIFIED CITY SCHOOL DIST.
	Instructional Assistant	" " "
1985–88	***Staff Student Nurse***	BRONX LEBANON HOSPITAL & CALVARY HOSPITAL, New York, NY
1983–84	***Adolescent Activity Counselor***	YOUTH CORPS PROGRAM, " "

EDUCATION

M.A., Clinical Psychology, 1996; John F. Kennedy University, Orinda, CA
Certificate in Chemical Dependency, JFKU, 1996
M.F.C.C. candidate, J.F.K.U. – anticipated April 1998
B.A., Anthropology/Spanish Culture, 1991; Drew University, Madison, NJ
B.A., Psychology, 1991 – United States International University, San Diego, CA
LPN (Vocational Nursing Diploma), 1986 – Morris School Of Nursing, New York

HONORS

- National Dean's List of American Universities and Colleges
- Selected candidate for the 1989-91 Annual Encyclopedia
- Who's Who Among Students in American Universities and Colleges
- Selected candidate for annual Encyclopedia, based on academic standings, community service, leadership ability and future potential, 1990-91
- New York Association of Nurses
- Merit scholarship award

PROFESSIONAL AFFILIATIONS

California Association of Marriage & Family Therapists
Association of Black Psychologists, Bay Area Chapter

EILEEN T. SCHULMAN
1412 California St., Apt. 206
San Francisco, CA 94109
(415) 262-3977

Objective: Social work position with University of California Medical Center.

PROFILE

- 10-year background in health and social services.
- Genuine concern for and sensitivity to clients.
- Excellent problem solving and counseling skills.
- Enjoy the challenge of assisting clients in time of crisis.
- Work well as member of an interdisciplinary health team.
- Thorough familiarity with S.F. community resources.

RELATED ACCOMPLISHMENTS

Counseling
- Co-developed a unique Bereavement Counseling Program for group counseling of individuals facing recent loss of a loved one.
- Served as practitioner in wide range of situations, utilizing skills in:
  – short/long-term counseling; – crisis intervention; – phone counseling;
  – family counseling; – bereavement counseling.
- Successfully engaged resistant clients in counseling.

Interviewing & Needs Assessment
- Conducted psycho/social assessments of individuals and families.
- Developed individualized needs assessments for homebound patients, providing:
  – case management; – counseling; – supervision of care attendants;
  – home care hours; – sickroom equipment; – home care supplies.
- Monitored home care plans for indigent homebound cancer patients.
- Provided comprehensive discharge planning, identifying adequate placement resources for hospitalized patients.

Administration
- Administered Home Health Program which won award for Service to the Community during my tenure.
- Co-authored a popular and effective bimonthly Newsletter for Home Care Attendants, providing career information, training opportunities and professional support.
- Co-developed Attendant Training Workshops, Rap Sessions and Annual Attendant Recognition Dinner.
- Prepared annual budget and biannual Program Report, including all statistical program information.
- Organized an information network and referral system on health care in the Bay Area to provide highest quality referrals.

– Continued –

EMPLOYMENT HISTORY

1994–present	**Program Coordinator**	MILLER-BUNTING PROGRAM of the American Cancer Society, S.F.
1993–94	**Program Social Worker**	" " "
1992–93	**Social Work Intern**	PACIFIC PRESBYTERIAN HOSPITAL, S.F.
1991–92	**Social Work Intern**	DEPT. of SOCIAL SERVICES, S.F., Children's Protective Services
1986–93	**Licensed Vocational Nurse**	ST. LUKE'S HOSPITAL, S.F. Adult Medical/Surgical; Pediatrics
1985–86	**Licensed Vocational Nurse**	VETERANS' ADMIN. HOSPITAL, S.F., Cardiology/Neurology

EDUCATION

MSW – San Francisco State University, 1993
Additional Training: Oncology Social Work, Memorial Sloane-Kettering Cancer Center, 1995
LVN – College of California Medical Affiliates, San Francisco, 1985

PROFESSIONAL AFFILIATIONS

National Association of Social Workers
Bay Area Social Workers in Health Care (Executive Committee 1993-95)
San Francisco Grief Care-Givers Network

BENJAMIN FARBER, MFCC

Licensed Psychotherapist
30566 Hillegass Avenue
Berkeley, CA 94705
(510) 909-7866

SUMMARY

- Extensive experience assessing and treating issues related to workplace relationships, family relationships, and stress disorders.
- Skill in working as member of a professional treatment team.
- Demonstrated effectiveness in issues of alcoholism and recovery.
- Specialist in developing self-worth and self-esteem.
- Able to instill confidence in a positive treatment outcome.

PROFESSIONAL EXPERIENCE

Alcoholism & Recovery

- Treated identified alcoholics, referred by alcohol treatment programs and other therapists, at early stages of recovery:
  - Assessed potential alcoholics in the workplace; recommended appropriate treatment.
  - Educated clients on the predictable course of early recovery, e.g. depression, loss of control;
  - Identified cues for potential relapse to design more effective coping styles related to their social support system, job-related stresses, and family relations.
- Effectively applied disease model of alcoholism, combining abstinence with adjunct treatment: e.g., Alcoholics Anonymous; family therapy.
- Diagnosed and treated co-alcoholics, adult and adolescent children of alcoholics.

Workplace Relationships

- Mediated the resolution of workplace problems with employees of school, government, military and industry:
  - Advised supervisors having difficulty asserting authority;
  - Empowered employees to successfully identify and effectively resolve conflict with supervisors and co-workers, teaching assertiveness and communication skills.

Stress Disorders/Stress Reduction

- Assessed and diagnosed burnout and other stress related disorders.
- Taught relaxation techniques, self-hypnosis, exercise programs, and development of self-worth, in the context of individual psychotherapy.

Couple & Family Relationships

- Assisted couples in:
  - Identifying their issues of conflict;
  - Articulating individual needs within family / couple context;
  - Developing skills in communicating anger or uncomfortable feelings, negotiating and problem solving;
  - Accepting differences as normal in healthy relationships.
- Successfully applied structural family therapy techniques to assist step-families as well as intact families in developing clearer and more cooperative generational roles.

– Continued –

EMPLOYMENT HISTORY

1993–present	**Co-director**	OAKLEY COUNSELING ASSOCIATES, Oakland & Berkeley
1990–93	**Psychotherapist**	RICHFIELD COUNSELING ASSOCIATES, Alameda
1988–90	**Psychotherapist**	FAMILY SERVICE AGENCY, Walnut Grove
1987–88	**Primary Therapist**	CROWN HOME, residential treatment facility, SF
1985–86	**Counselor/**	PARKHURST PLACE INC., psychiatric halfway house, SF "
"	**Treatment Coord.**	" " "

CONSULTANT & TRAINER EXPERIENCE

1994–95	**Supervisor**	Graduate Training Program, Bay Graduate School of Marital & Family Therapy, Oakland
1990–93	**Chairman**	School Attendance Review Board, Oakley School Dist.
1989–93	**Consultant**	Special Education Dept., Berkeley High School
1989–92	**Consulting Instructor**	Greer College, Hayward

PROFESSIONAL EDUCATION & DEVELOPMENT

Relevant Specialized Training

Alcoholism Treatment, 1991–present, seminars with:
- Matthew Skinner, M.D.
- Corrine Williamson, Ph.D.
- Bay Graduate School of Marital & Family Therapy

Group Therapy Training
- Gestalt Institute of New York, 1978–80
- Gestalt Institute of San Francisco, 1983–84

Marriage & Family Therapy
Bay Graduate School of Marital & Family Therapy
- Karl Greenbaum, Ph.D
- Trevor Howard, M.D.
- Leroy Russ, M.D.
- Jane Holland, M.A.

Psychotherapy Training
- Ego Psychology / Object Relations Theory & Techniques
- Solano Center for Psychological Services – Albany
- Control Mastery Theory & Techniques
- Roger Pritchard, M.D.

Hypnotherapy, Stress Reduction & Meditation Techniques
- UC Berkeley and George Solis, Ph.D.
- Yoga Societies of New York & San Francisco

Professional Credentials

Licensed Marriage, Family and Child Counselor, MF987654
Community College Instructor Credential, California No. YS 1234
Hypnosis Certification No. 99299

Health Resumes

Margo Segall	Medical investigator/counselor	143
Julia Millhouse	Manager, geriatric care facility	144
Harriet Bloom	Health educator/nutritionist	146
Harriet Bloom	Project coordinator, special study	148
Robert Adminster	Program director/gerontology	150
Ken Chesak	Body therapist/affiliate	151
Patricia Raines	Stress reduction specialist	152
Richard Jennings	EMT field supervisor	154
Pauline Masterson	Nursing services director	156
Tony Foote	Physical therapy aide	157
Carole Loomis	Medical/nursing/consultant	158

Margo got the job, over hundreds of other applicants.

MARGO SEGALL, R.N.

2131 Blackhawk Rd. • Berkeley, CA 94707 • (510) 275-2347

Objective: Medical investigator/counselor with Irwin Memorial Blood Bank.

SUMMARY OF QUALIFICATIONS

- 11 years' experience as a research clinical nurse.
- Thorough understanding of the protocols of human research and data collection.
- Skill in dealing with sensitive populations in a professional and concerned manner.
- Able to work independently and as a cooperative team member.
- Experienced and competent phlebotomist.

PROFESSIONAL EXPERIENCE

COUNSELING

- As a consultant with the Berkeley Unified School District, conducted crisis intervention and long-term counseling with individuals of diverse backgrounds on issues of confinement, illness, and institutional group living.
- Advised prospective research volunteers of positive test results indicating presence of venereal disease, TB, high blood pressure or abnormal blood values.
- Directed disqualified volunteers to appropriate referral agencies as necessary for medical follow-up.

PHLEBOTOMY

- Drew blood samples for high volume health screenings in over 40 human nutrition research studies.
- Performed venipunctures and / or IV's throughout each study, as specified by research protocol.
- Prepared blood samples for analysis or transport.

MANAGEMENT/SUPERVISION

- Served as head nurse over 6-month period, supervising support staff, research volunteers and graduate students, for a UC Berkeley nutrition-related study.
- Authored procedure on blood / body fluid precautions, and delivered in-service training talk to staff.
- Taught data collection techniques to research participants.
- Assembled final report data at conclusion of studies.

KNOWLEDGE OF AIDS

- Completed three continuing education workshops on AIDS:
  "AIDS: Fears, Facts, and Fantasies," Letterman Hospital and Shanti Project;
  "AIDS: The Spectrum of the Acquired Immune Deficiency Syndrome," Letterman Hospital;
  "AIDS: Infection Control for Health Workers," San Francisco Dept. of Public Health.
- Familiar with "Guidelines for HLV Antibody Testing in the Community," as provided by the State of California.

EMPLOYMENT HISTORY

1984–present	***Research/Clinical Nurse II***	Univ. of California / USDA, Berkeley / SF
1990–94	***Health Consultant***	Berkeley Unified School Dist. (concurrent w / above)
1976–89	***Teacher, Special Education***	Richmond Unified School District

EDUCATION & TRAINING

B.A., English & Art – UC Berkeley, 1974; Teaching Credential, 1975
R.N. – MERRITT COLLEGE, 1981; CPR Certified, 1985

HEALTH

JULIA MILLHOUSE

68809 Steiner Street, San Francisco, CA 94117
(415) 302-6113 home (415) 815-6609 messages

Objective: Middle management position in a geriatric extended care facility, involving staff development, education, and direct supervision of patient caregivers.

SUMMARY OF QUALIFICATIONS

- Clinical expertise and experience in all aspects of geriatric care.
- Committed to highest standards of care for the geriatric population.
- Proven skill in teaching and supervising professional nurses.
- High energy leader who inspires and challenges others to excellence.

RELEVANT EXPERIENCE

Supervision

- Coordinated and supervised nursing team responsible for 24-hour care of geriatric patients, at King Medical Center Nursing Home Care Unit:
  - Expanded program of preventive care, minimizing need for hospitalization;
  - Upgraded anecdotal reporting on nursing care, assuring that reports were adequate for judging staff advancement potential;
  - Provided personal support of staff in bettering themselves through education;
  - Assured that family members were very well informed and involved.
- Successfully filled in as charge nurse during the absence of head nurse.

Teaching

- Taught in-service programs for floor staff on geriatric care issues:
  ...diabetes ...psycho-social care needs ...infection control ...physiology of aging.
- Facilitated Wives' Support Group discussion on current trends in the care of diabetes.
- As Teaching Assistant for expert in Geriatric Care Delivery at Livermore General Hospital:
  - Wrote and graded exams;
  - Arranged for guest speakers;
  - Prepared syllabus; selected reading lists;
  - Advised students on academic and clinical issues.

Planning/Policy

- Co-chaired joint committee on Infection Control and Product Evaluation, involving nurses from throughout hospital, responsible for assessing staff adherence to policy standards:
  - Effectively represented the unique interests and priorities of the nursing home staff;
  - Voluntarily continued as an active committee member while part-time employed.
- Took the initiative to develop a plan responding to pending federal legislation, showing how computers could be used in developing mandatory multi-disciplinary care plans; currently involved in the on-going development of this plan.

– Continued –

EMPLOYMENT HISTORY

1994–present	***Staff RN***	KING MEDICAL CENTER Nursing Home Care Unit, Livermore, CA
1994–95	***Teaching Assistant***	SCHOOL OF NURSING, Livermore General Hospital
1993–94	***Staff RN/Team Leader***	ST. GEORGE MEDICAL CENTER, Livermore, CA
1992–93	***Staff RN***	DANE MEMORIAL MEDICAL CENTER, Reno, NV
1991–92	***Full-time student***	Univ. of Nevada, Reno, Orvis School of Nursing
1991 (summer)	***Student Nurse Tech.***	DANE MEMORIAL MEDICAL CENTER, Reno, NV
1984–90	***Resort Supv./Waitress***	SO. LAKE TAHOE RESORTS, NV
1979–80	***Teaching Assistant***	SO. LAKE TAHOE UNIFIED SCHOOL DISTRICT, CA

EDUCATION

B.S., Nursing – UNIVERSITY OF NEVADA, RENO, 1992
B.A., History – UNIVERSITY OF CALIFORNIA, LOS ANGELES

HARRIET BLOOM
5720 Hillegass Ave.
Berkeley, CA 94705
(510) 622-5917

Objective: Position as Health Educator/Nutritionist.

SUMMARY

- Over ten years' professional experience in health education.
- Proven successful in managing simultaneous projects.
- Talent in designing systems for smooth program operation.
- Skilled teacher and trainer; able to inspire others.
- Strong commitment to promoting wellness and preventing disease.

PROFESSIONAL EXPERIENCE

Administration

- Revamped and managed the Health Education Program of HealthAmerica, restoring and expanding programs that had been neglected.
- Developed administrative systems to handle all details for implementing programs, for both the federal and state health education departments at HealthAmerica.
- Won federal funding for a unique family program, integrating medical care with a supplemental food program, maximizing the effectiveness of the programs.

Supervision & Training

- Trained paraprofessionals in basic nutrition concepts and group leadership skills.
- Supervised nutrition program staff of 3; supervised Health Education Assistant.
- Served as preceptor for UC Berkeley and SF State graduate students.
- Coordinated in-service programs for physicians:
  ...smoking cessation counseling; ...latest developments in the management of obesity.

Nutrition Counseling & Health Promotion

- Counseled a very wide range of patients in preventive nutrition and chronic disease.
- Developed an effective approach to weight management, matching patients to programs with highest probability of success.
- Promoted self-care classes to increase patients' knowledge and skills and their involvement in health care decisions.

Program Development

- Developed and implemented both patient and practitioner surveys to assess health education needs.
- Evaluated patient education materials, adapting and altering as needed.
- Developed a variety of evaluation tools to assess patient knowledge and behavior change.

Public Relations & Community Liaison

- Collaborated on design of brochure and fliers for health education classes and programs.
- Organized and chaired a community-based Health Advisory Committee for Alameda Head Start.
- Set up a monthly Health Education Advisory Committee at HealthAmerica.

– Continued –

See Harriet's other resume on page 148.

EMPLOYMENT HISTORY

1995–present	**Health Education Coordinator**	HEALTHAMERICA, Oakland
1990–95	**Nutrition Project Director**	HEALTHAMERICA, Oakland
1988–present	**Senior Nutritionist**	HEALTHAMERICA, Oakland
1989–91	**Nutrition Consultant** (concurrent with above)	WESTINGHOUSE HEALTH SYSTEMS, San Francisco
1986–88	**Health/Nutrition Coordinator**	ALAMEDA HEAD START, Alameda
1984–86	**Nutrition Consultant**	DEVELOPMENT ASSOCIATES, SF

EDUCATION & CREDENTIALS

M.P.H., Nutrition – University of California, Berkeley – School of Public Health
Dietetics course work, University of California, Berkeley
B.A., Psychology – Vassar College, Poughkeepsie NY

Registered Dietitian (R.D.) 1986
California Community College Instructor Credential

HARRIET BLOOM

5720 Hillegass Ave.• Berkeley, CA 94705 • (415) 622-5917

Objective: Position with Highland Hospital as Project Coordinator, Breast Feeding Support Project Study

SUMMARY OF QUALIFICATIONS

- Experienced in developing systems essential to the smooth operation of a program.
- Proven successful in managing simultaneous projects.
- Skilled teacher and trainer, able to inspire others.
- Excellent counseling and interviewing skills.
- Strong commitment to promoting the benefits of breast feeding.

PROFESSIONAL EXPERIENCE

ADMINISTRATION

- Revamped and managed the Health Education Program of HealthAmerica, restoring and expanding programs that had been neglected.
- Developed administrative systems to handle all details for implementing program for both the health education department and the WIC program at HealthAmerica.
- Won federal funding for a unique WIC program, integrating medical care with a supplemental food program, maximizing the effectiveness of the programs.
- Developed an effective method for highlighting entries in medical records to assure timely exchange of important information between doctors and WIC program staff.

SUPERVISION & TRAINING

- Trained paraprofessional Nutrition Education Assistant in basic nutrition concepts and group leadership skills.
- Trained paraprofessionals (family advocates) in a home-based Head Start Program in basic nutrition and child development.
- Supervised WIC staff of 3; supervised Health Education Assistant.
- Served as preceptor for UC Berkeley and SF State graduate students.

BREAST FEEDING PROMOTION

- Initiated a successful program which doubled the incidence of breast feeding in a low income minority population (WIC):
  - greatly increased acceptance by identifying barriers (concerns, fears, myths);
  - added group education and individual counseling to program requirements;
  - developed interdisciplinary team immediately available to resolve early problems.

NUTRITION COUNSELING

- Counseled a very wide range of patients in nutrition:
  - prenatal - infant and pediatric - adolescent - mid-life - geriatric

PUBLIC RELATIONS & COMMUNITY LIAISON

- Organized and chaired a community-based Health Advisory Committee for Alameda Head Start.
- Set up a monthly Health Education Advisory Committee at HealthAmerica

– Continued –

See Harriet's other resume on page 146.

HARRIET BLOOM
Page two

EMPLOYMENT HISTORY

1995–present	**Health Education Coordinator**	HEALTHAMERICA, Oakland
1990–95	**Nutrition Project Director**	HEALTHAMERICA, Oakland
1988–present	**Senior Nutritionist**	HEALTHAMERICA, Oakland
1989–91	**Nutrition Consultant** (concurrent with above)	WESTINGHOUSE HEALTH SYSTEMS, San Francisco
1986–88	**Health/Nutrition Coordinator**	ALAMEDA HEAD START, Alameda
1984–86	**Nutrition Consultant**	DEVELOPMENT ASSOCIATES, SF

EDUCATION & CREDENTIALS

M.P.H., Nutrition – University of California, Berkeley
School of Public Health Dietetics course work, University of California, Berkeley
B.A., Psychology – Vassar College, Poughkeepsie NY

Registered Dietitian (R.D.) 1986
California Community College Instructor Credential

Robert L. Adminster

334 Marguerite Drive
Cleveland, OH 44163
(123) 456-7899

OBJECTIVE

Position as Program Director, involving planning, management, & marketing.

SUMMARY OF QUALIFICATIONS

- Fifteen years' experience in management of senior health care facilities.
- M.A., Gerontology, and Nursing Home Administrator's License, Ohio.
- Experience directing complex projects from concept to fully operational status.
- Solid understanding of marketing and sales strategies in health care.
- Able to work effectively in a variety of organizational structures.

EXPERIENCE

1994–present

OHIO BAPTIST HOMES, Cleveland, OH
Vice President of Operations
Oversaw operations of a retirement center and two skilled nursing facilities, serving over 450 residents.
- Achieved the facilities' best year in recent history, financially and by census.
- Developed uniform standards in personnel policies and procedures.
- Devised a medical records system that is easily usable and meets regulations.

1991–94

WALNUT HILL HOMES, Mirabelle, OH
Administrator
Administered this full-service 173-unit retirement center employing nearly 100 staff, from pre-construction to full operation.
- Developed marketing programs to secure the project's approval and funding:
  - Obtained Certificate of Need for the skilled nursing facility.
  - Directed the facility from a staff of one to full operations and full staffing.

1989–91

FLORIDA ASSOCIATION OF HOMES FOR THE AGING, Pensacola FL
Director of Membership Services
Directed both membership recruitment/retention and professional education programs for this statewide association of not-for-profit service providers.
- Successfully broadened the membership base through enhanced services.
- Designed and launched an improved calendar of educational workshops.
- Introduced a successful marketing program to build membership.

1986–88

BLOOMINGTON MANOR, Bloomington, PA
Administrator
Managed day-to-day operations of a 44-bed skilled nursing facility, including transition from family-run company to multi-facility corporate ownership.
- Implemented marketing, PR, and sales programs to aid expansion:
  - Initiated a meals-on-wheels service that improved community relations.
  - Brought the facility up to date and into regulatory compliance.

1984–86

ADMINISTRATION ON AGING, Washington, D.C.
(U.S. Department of Health and Human Services)
Program Specialist
Researched programs for the elderly offered by federal, state, and local agencies.
- Authored statements used by federal officials and members of Congress.
- Conferred with and advised numerous agencies involved with the elderly.

EDUCATION

M.A., Gerontology, 1984 – University of Maine, Portland ME
Nursing Home Administrator's License, Ohio

REFERENCES

Available on request.

KEN CHESAK

INTEGRATED EDUCATIONAL BODY THERAPY
98576 Telegraph Avenue
Berkeley, CA 94705
(510) 234-5678

Objective

Professional affiliation with a health practitioner, health care center or related service, providing body therapy and/or educational services, in-house or out-call referrals.

Summary of Qualifications

- Advanced certification in educational body therapy from the Institute for Educational Therapy.
- Certification in massage by National Holistic Institute.
- Commitment to professional development.
- Sensitivity and responsiveness to the needs of clients.
- Self-motivated; experience in managing a successful practice.

Professional Experience and Approach

EDUCATIONAL BODY THERAPY

- I apply an educational approach to helping clients with mind/body heath issues, increasing their self-help skills and their cause-and-effect awareness, using:
  ...deep tissue massage ...peripheral joint articulation ...spinal articulation
  ...movement therapy ...re-patterning restrictive and inefficient movement habits
  ...postural and structural alignment through body awareness; ...balance of tone
  ...stress release through awareness of breath.

BODYWORK/MASSAGE

- Worked with hundreds of individual clients in private practice, applying a wide range of therapeutic techniques: ...Swedish ...Acupressure ...Polarity ...Sports Massage

Employment History

1995–present	***Massage Therapist***	PRIVATE PRACTICE, Berkeley
1990–95	***Banquet Waiter***	OAKLAND SERVICE CO., caterers, Oakland
1990–95	***Music Instructor***	APEX MUSIC, Oakland; and B&A MUSIC, Pt. Richmond

Relevant Education & Training

- INSTITUTE FOR EDUCATIONAL THERAPY, Berkeley, 200-hour Body Therapy training, 1995; Instructor, Jim Tierra
- NATIONAL HOLISTIC INSTITUTE, Oakland, 100-hour Massage Certification, 1995
- Polarity Therapy Center of San Francisco, 40-hour Polarity Therapy training, 1995
- Sports Massage Workshop, 30 hours, Oakland, 1996; Instructor, Teri Tripp
- Developmental Movement and Experiential Anatomy; Edith Jay, Sausalito, 1995
- Movement Workshop, Bartenieff Fundamentals, 5th St. Studios; Betsy Torrey, Berkeley, 1995
- Movement and Sound Workshops; Nina Rowe and Bob Arvey, instructors, 1992–93
- Skinner Releasing Techniques; Contact Improvisation (Movement Therapies) 5th St. Studios; Robert George and Catherine Martin, instructors, 1995

PATRICIA RAINES, B.S., R.N.

Specialist in Stress Reduction & Biofeedback

960 Los Cerros Avenue
Alameda, CA 94598
(510) 111-5760

PROFILE

- Demonstrated success in helping clients resolve the physical and emotional symptoms of mishandled stress.
- Personal and professional commitment to the empowerment of clients; creative, intuitive approach, allowing them to tap into their own solutions.
- Skill in providing a safe, supportive, and nonjudgmental environment.
- Unique background combining 15 years in nursing, a degree in psychology, and 8 years in motivational management.

PROFESSIONAL EXPERIENCE

Individual Consulting & Philosophy

Effectively consulted with hundreds of individuals, facilitating them in discovering new solutions to immediate or chronic upsets, applying these proven principles:

- Created a safe, supportive, attentive and nonjudgemental atmosphere.
- Identified the issue and the immediate goal for the session.
- Encouraged them to explore feelings, as opposed to thoughts about the issue.
- Kept them on track; persistently restored the focus; surrendered my own judgements.
- Validated and acknowledged them in a loving atmosphere that allowed them to look within.
- Supported them in going beyond what they previously thought possible, to find their own answers within themselves.

Teaching/Motivational Training

Served as program coordinator, trainer and leader with Motivational Management Service:

- Trained and coordinated Workshop Facilitation Teams for the duration of 3-month periods; training them to work one-to-one with participants and support the workshop leaders.
- Co-led workshops on self-esteem; led the follow-up 6-week Integration Support Series.
- Led Self-Empowerment, a 2-month series of weekly meetings.
- Coordinated 3-month training series, Consultants Training, assisting participants to define and actualize goals.

Relevant Medical Experience

- Developed a broad understanding of medical conditions and illnesses, as nurse in all units at Alta Bates Hospital: ...Medical ...Surgical ...Orthopedic ...Neurology ...Oncology ...Obstetrics & Gynecology ...Intensive Care ...Coronary Care.
- Designed a comprehensive in-service training program for nurses at St. Joseph's Hospital.
- Coordinated the intervention services of the Department of Social Services, Office of Mental Health, community agencies and health care professionals, to assure appropriateness to students' needs.

– Continued –

Patricia indicates that she's looking for a professional affiliation rather than employee status in the way she presents herself at the top of the resume.

EMPLOYMENT HISTORY

1986–present	***Staff Nurse***	ALTA BATES HOSPITAL, Berkeley
1975–85	Family management	
1972–74	***Nursing Supervisor***	UNIV. of CALIFORNIA MEDICAL CENTER , S.F.
1971–72	***Asst. Dir. of Nursing***	ST. JOSEPH'S HOSPITAL, Denver, CO
1969–71	***Head Nurse***	PRESBYTERIAN HOSPITAL, New York, N.Y.
1968–69	***Staff Nurse***	PRESBYTERIAN HOSPITAL, New York, N.Y.

EDUCATION & TRAINING

B.A., Psychology – WILSON COLLEGE, Chambersburg, PA
B.S., R.N., COLUMBIA UNIVERSITY, New York, NY

Motivational Management Training:
- Group Consulting, Mediation, and Partnership
- Leadership Training Programs
- Self-esteem Workshop
- Consultants Training

Biofeedback Institute, San Francisco

RICHARD JENNINGS

8009 Mountain Blvd. • Oakland, CA 94602
(510) 234-5678

Objective: County Field Supervisor, Emergency Medical Services.

SUMMARY

- Three years' experience as a paramedic, five years as an EMT, and one year's experience in staff supervision.
- Proven ability to respond immediately and confidently in emergencies.
- Able to function at top performance throughout a 24-hour shift.
- Mature lifestyle, compatible with emergency work.
- Excellent relations with the public and the community.

EXPERIENCE

1995–present **Paramedic Crew Chief** BAY MEDICAL SERVICES, Berkeley/Oakland

Supervision

- Led pre-hospital-care team, as Crew Chief for BMS, serving as liaison between paramedics, hospitals and fire department:
  – Provided monthly training updates for the firemen I worked with;
  – Mediated communication problems among professionals;
  – Provided follow-up medical information on cases jointly handled.

Training & Quality Assurance

- Selected for and served on Paramedic Peer Review Committee at Eden Hospital, monitoring paramedic response in order to improve treatment of patients.
- Trained new paramedics in the field portion of their state-required training time, focusing on communication and decision-making skills in medical emergencies.
- Wrote draft of training manual for new medics at Bay Medical Services.
- Helped upgrade the skills of weak paramedics, providing evaluation and customized retraining.

1994 **Paramedic** KING AMERICAN AMBULANCE, San Francisco

Crisis Evaluation & Response

- Effectively evaluated thousands of emergencies.
- Adapted immediately to changing circumstances in medical emergencies, setting priorities and constantly reevaluating them.

Community Relations

- Educated the public on the role of emergency medical services, through demonstrations, lectures and small group talks.
- Taught public CPR classes; served as volunteer medic for public events; trained nurses and hospital employees in trauma skills at Eden Hospital.

– Continued –

Compare this with Richard's other resume on page 243.
See Richard's cover letter on page 266.

RICHARD JENNINGS
Page two

1992–93 **EMT-1A** ALLIED AMBULANCE, Oakland

- Developed a high level of skill in medical emergency techniques:
  ...patient assessment ...taking blood pressure and pulses
  ...wound management ...splinting fractures
  ...applying oxygen ...CPR and advanced life support

1987–91 **Warehouseman/Driver** GOOD GUYS; SAUSALITO DESIGN; PACIFIC FLOORING retail businesses, SF Bay

- Supervised warehouse staff at Sausalito Design:
  – Trained and guided personnel;
  – Monitored safe and accurate filling of orders and thoroughness of paperwork;
  – Researched customer complaints to identify error and resolve problems.

1984–86 **Day Care Driver** EASTER SEAL SOCIETY, San Francisco

EDUCATION & SPECIALIZED TRAINING

B.A., Sociology – SAN FRANCISCO STATE UNIVERSITY
A.A., Criminology – SAN JOAQUIN DELTA COLLEGE, Stockton

Paramedic Training; EMT-1A Training – City College of San Francisco
Basic Life Support & Advanced Cardiac Life Support Certificates
– American Heart Association.
Ambulance Driver's License – Advanced Airway Management Training
Additional courses in Anatomy, Physiology, Chemistry, 1995-96

PAULINE MASTERSON, MPA/HSA

457 Hathorn Court, San Francisco, CA 94117
(415) 204-6906

OBJECTIVE: **Position as Director of Nursing Services** for a hospital or HMO.

SUMMARY OF QUALIFICATIONS

- MPA degree and over ten years' experience in nursing.
- Special strength in promoting an atmosphere of professionalism.
- Committed, enthusiastic, people-oriented leader.
- Highly successful in developing collaborative practice.
- Concerned with balancing quality nursing and cost containment.

RELEVANT EXPERIENCE

ADMINISTRATION/MANAGEMENT

- Managed a 45-bed medical teaching unit on a 24-hour / 7-day basis, including long / short-term planning, hiring, interviewing and orienting new staff.
- Developed a management style emphasizing professional responsibility:
  - Delegated authority for problem solving; – Involved staff in decision making;
  - Held staff accountable for their actions; – Helped staff to resolve personal issues.
- Raised Quality Assurance Scores by incorporating staff input into planning process.
- Developed and implemented cost containment proposals, for example:
  - Decreased staff sick time use by involving them directly in the scheduling;
  - Informed staff of materials costs and encouraged ideas for economizing.
- Analyzed fiscal trends in terms of budget goals.

STAFF DEVELOPMENT

- Encouraged creativity and professionalism of the nursing staff, resulting in their taking the initiative in forming a unique self-governance Nursing Practice Committee.
- Upgraded the employee performance evaluation process to incorporate criteria reflecting increased individual accountability in nursing.

PROMOTING PROFESSIONALISM

- Expanded and strongly supported high school student apprenticeship programs by authorizing one-on-one preceptorships with staff nurses.
- Provided staff nurses as preceptors for student nurses during clinical rotation.
- Serve as active bureau speaker and CPR instructor for American Heart Association.
- Conducted research, and incorporated findings, on perceptions of leadership styles and their relationship to professional autonomy and accountability of staff nurses.
- Served as on-site investigator for Stanford Research Consortium.

EMPLOYMENT HISTORY

1994–present	**Nurse Manager**	MEMORIAL HOSPITAL & MEDICAL CENTER, S.F.
1991–94	**Clinical Care Coord.**	MEMORIAL HOSPITAL & MEDICAL CENTER, S.F.
1986–91	**Staff Nurse II**	MEMORIAL HOSPITAL & MEDICAL CENTER, S.F.
1983–86	**Staff Nurse I & II**	SANTA CLARA VALLEY MEDICAL CENTER, San Jose

EDUCATION & LICENSURE

MPA / Health Services Administration – UNIVERSITY OF SAN FRANCISCO, 1996
BS, Nursing – CALIFORNIA STATE UNIVERSITY, SAN JOSE, 1983
Registered Nurse, State of California

TONY FOOTE

1992-C Parker Street
Berkeley, CA 94704
(510) 399-0102

Objective: Position as Physical Therapy Aide.

PROFILE

- Compassionate, professional approach and commitment to service-oriented work.
- Experienced and effective in assisting people with medical disabilities.
- Interest and knowledge in the field of physical therapy.
- Long-term goal to practice as a registered physical therapist.
- Degree in biology; course work in human anatomy and human physiology.

EMPLOYMENT HISTORY

1994–present **Assistant Teacher** HARWOOD DAY SCHOOL, Oakland

- Assisted with speech therapy in a classroom for language-impaired children, both as a volunteer and as a paid employee.
- Formulated the speech therapy program for a language-delayed student under direction of a speech therapist.

1992–94 **Personal Care Attendant** SATELLITE SENIOR HOMES, Oakland

- Maintained records of medicines taken, and relevant observations of physically handicapped and frail elderly residents, in this senior retirement facility.
- Implemented adaptive aquatic routines with a physically handicapped adult.

1990–91 **Instructional Assistant** PUBLIC SCHOOLS, Eugene & Springfield, OR

- Assisted in supervising recreation activities of mentally handicapped children and adults, as volunteer at city-sponsored specialized recreation program.
- Instructed basic academic subjects in classroom for learning-delayed children.
- Developed behavioral and academic strategies, and recorded academic performance, for emotionally handicapped adolescents.

1988–89 **Developmental Aide** EASTER SEAL SCHOOL, Eugene, OR

- Served as a physical therapy aide, working with seriously handicapped children:
  - Performed basic physical therapy routines to facilitate physical development;
  - Positioned children to minimize pathology.

1987 **Bus Driver** PEARL BUCK SCHOOL for mentally handicapped, Eugene, OR

EDUCATION & TRAINING

B.A., Biology – UNIVERSITY OF OREGON, Eugene, OR
– Human anatomy, human physiology, massage, psychology, abnormal psychology, language acquisition, statistics.

CAROLE R. LOOMIS

MEDICAL / NURSING CONSULTANT
1490 Lincoln Blvd.
Boulder, CO 80306
(123) 456-7890

SUMMARY

- Masters degree in cross-cultural international nursing and administration.
- Hands-on experience in virtually all areas of the medical industry.
- Broad perspective of people based on cross-cultural education, participation in an international program, and travel.
- Successful experience with refugee nurses, providing nurse training, US cultural orientation, and job development.
- Knowledge of computer applications in the medical field.

RELEVANT EXPERIENCE

REFUGEE NURSE TRAINING – JOB PLACEMENT

As Program Director for **Refugee Nursing Program** at Burns College, Denver:
- **Taught** cross-cultural Health Care and Nursing classes.
- **Counseled** students on **cultural adjustment**; advised them on **academic requirements.**
- Administered the program.
- **Developed nursing jobs** within the community; secured **job placement** of students.

EXPERTISE IN MEDICAL / HEALTH FIELD

- Staff Nurse for over 10 years, working with a wide range of medical areas:

  ...Acute Medical and Surgical conditions ...Intensive Care
  ...Pediatrics ...Geriatrics ...Neurology
  ...Burns and Reconstructive Surgery ...Orthopedics

- Served as Red Cross **Health Educator; Regional Coordinator** for Genetic Screening Program; and **Health Services Researcher.**

BUSINESS ADMINISTRATION

- Successfully sold a **medical** billing service, conducting telephone cold-calls followed by in-person sales presentations at physicians' offices.
- Completed Dale Carnegie course, **Effective Speaking and Human Relations,** including principles applicable to effective presentation, sales and marketing.

PROMOTION

- Successfully **promoted to clinics and physicians** a program of genetic screening, gaining acceptance through persuasion and program explanation.
- Promoted physician's medical practice and state-of-the-art services, arranging for guest speaker engagements at hospitals and community organizations.

- Continued on page two -

With careful planning, even a complicated work history can be made clear and periods of unemployment can be fitted neatly into the chronology.

WORK HISTORY

1996–present	**Independent Consultant**	Job development for US placement of overseas nurses
1994–95	**Medical Sales Consultant**	MEDICO DATA, Denver; and UROLOGICAL ASSOC. MEDICAL GROUP
1993–94	**Health Care Program Coor.**	METROPOLITAN HOSPITAL, Denver
1992–94	**Graduate student**	UNIVERSITY OF COLORADO, Denver
	...Compliance Nurse	DENVER GENERAL HOSPITAL
	...Coordinator	REFUGEE NURSING PROJECT
1988–91	**Family / maternity leave**	
1985–87	**Staff Nurse**	STUDENT HEALTH SERVICE, Univ. of Colorado
1982–84	**Full-time student**	UNIVERSITY OF COLORADO, Denver
	...Health Educator	RED CROSS
	...Staff Nurse	UNIVERSITY OF COLORADO, Denver

EDUCATION & TRAINING

M.S., Cross-Cultural International Nursing / Administration
B.S., Nursing; Public Health Nurse Certificate, University of Colorado, Denver
R.N., Kings General Hospital, London, England

Additional Course Work

•Marketing •Economics •International Relations •Dale Carnegie Course

Computer Applications

Basic computer training, University of Colorado School of Nursing.
Software training in data transfer, with Genetic Screening program.
Computer training in medical billing service.

REFERENCES

Available on request

Marketing Resumes

Tracy Hazelton	Market research/analysis	161
Andrea Hughes	Public Relations/marketing	162
Claudia Giselle	Marketing/PR/promotion	163
Cynthia Mayer	Marketing/sales/client services	164
Elizabeth Julian	Publicist/arts organization	165
Fran Morgan	Sales/marketing, PR, promotion	166
Gary Bradley	Marketing, sales management	168
Gary Rosekrans	Accounts executive, ad agency	170
Gerald Davis	Sales and marketing	171
Betsy Emory	Project director, PR officer	172
Jean Bogart	Client services rep, medical center	173
Joanne Simpson	Market research, entry level	174
Joanne Fine	International marketing	175
Elizabeth Leonard	Public relations/promotion	176
Leslie Bowman	Fashion special events coordinator	177
Linda MacKinnon	Client services rep, health services	178
Sandra Cerrito	Marketing manager, medical products	179

TRACY HAZELTON

2699 Polk Street, Apt. 708
San Francisco, CA 94109
(415) 390-4223

OBJECTIVE

Market research position in consumer products/manufacturing, focusing on consumer demographics, market surveys, and needs/trends analysis.

PROFILE

- 20 years' experience in marketing; background in research.
- Highly motivated and dependable in achieving set goals.
- Talent for making creative ideas successful and profitable.
- Consistently take the initiative to solve problems.
- Extensive experience with PC's and spreadsheet software.
- Strong organizational skills; attention to detail.

— As Research Analyst and Marketing Coordinator for Pacific Tour —

Trends Analysis/Strategy & Planning

- Projected sales potentials by analyzing sales figures from current and past years.
- Gathered pricing and scheduling information on competitors and reported on changes needed to maintain a competitive edge.
- Increased the efficiency of Pacific Tours direct mail promotions by creating a program to consolidate passenger information, allowing for targeted mailings.
- Identified areas of customer satisfaction and motivation for repeat business, and successfully initiated changes to improve marketing program accordingly.

Demographics/Data Collection

- Input consumer data generated from questionnaire mailings and on-site surveys of Pacific Tour passengers.
- Assembled and input market data on potential Pacific customers having appropriate demographic characteristics.

Computer Software Skills

- Charted and graphed wide range of data in Lotus 1-2-3, including:
  ...Pacific Tour customer surveys and questionnaires ...Sales demographics.
- Composed correspondence using WordPerfect for Windows.

EMPLOYMENT HISTORY

1995–present	**Research Analyst Asst.**	PACIFIC TOUR LINES, San Francisco
1994–95	**Asst. Coordinator/Mktg.**	PACIFIC TOUR LINES, San Francisco
1989–93	**Training Program Coord.**	UNIV. OF FLORIDA Hospital & Clinics
1985–88	**Research Assistant**	UNIV. OF FLORIDA Marine Inst., Sapelo Is.

EDUCATION

B.A., Biology & Chemistry – CAL STATE UNIVERSITY, SONOMA
Graduate School, Biology – WESTERN WASHINGTON STATE COLLEGE

ANDREA HUGHES
3945 - 20th Street • San Francisco, CA 94114
(415) 628-0108 (office) • (415) 271-0055 (home)

Objective: Representative in a public relations/marketing department or agency.

SUMMARY

- Three years' experience in public relations and marketing.
- Highly competitive, and thrive in challenging situations.
- Maintain a sense of humor under pressure.
- Extremely sharp at quickly assessing needs and priorities.
- Diplomatic and assertive in dealing with people.

PROFESSIONAL EXPERIENCE

Public Relations

- Promoted Spanish products and tourism to Spain:
  - Organized conferences and group programs for traveling professionals;
  - Participated in trade shows, informing media representatives and members of the travel industry on Spain's travel offerings;
  - Advised media representatives and members of the travel industry on tourist attractions, cultural activities, living conditions, and political climate;
  - Wrote press releases directed to West Coast media on up-coming travel industry conventions.
- Developed contacts in Spain for American media producing features for US audiences; also coordinated their travel arrangements to, and within, Spain.

Marketing & Sales

- Successfully made cold-calls to most Bay Area real estate developers, generating new business for local architectural firm.
- Wrote proposals and assembled information packages describing architectural services for potential construction projects.
- Developed graphic design and coordinated production of firm's promotional brochures.
- Advised boutique retail customers on fashion and merchandise selection.

Organization/Administrative

- Developed ideas for creating new business, prioritized work projects, designed and implemented follow-up procedures, resulting in more efficient and profitable work flow.

EMPLOYMENT HISTORY

1992–96	***Public Relations Rep***	NATIONAL TOURIST OFFICE OF SPAIN, SF
1991	***Marketing Coordinator***	FEE & MUNSON Architects, San Francisco
1990	***Administrative Asst.***	CROWLEY MARITIME CORP., San Francisco
1988–89	***Salesperson*** (summers)	DAYTIME DELIGHT clothing store, Martha's Vineyard

EDUCATION

B.A., Drew University; Madison, NJ, 1989 – Anthropology/Spanish Culture
Studies in Anthropology, Universidad de los Andes; Bogota, Colombia, 1988
Extensive travel in Western Europe, North Africa and South America
Fluent in Spanish

CLAUDIA GISELLE

5002 - 45th Avenue
San Francisco, CA 94116
(415) 699-4104

Objective: Position in marketing, PR and/or promotions.

SUMMARY

- A born promoter, able to generate enthusiasm in others.
- Proven successful in increasing sales and customer base.
- 10 years' experience in public relations and promotions.
- Extremely well organized; follow through to the last detail.
- Committed to producing results above and beyond what's expected.

RELEVANT EXPERIENCE

Marketing • Public Relations

- Developed customer service procedures and training program for managers and staff.
- Originated and implemented marketing strategies to bolster sales at unprofitable store locations.
- Set up an advertising department for a restaurant in Eugene, Oregon, successfully establishing its new image as a community cultural center.
- Designed a successful marketing / PR department for Bay Area restaurant franchise.
- Established and maintained cooperative working relations with radio and print media, resulting in free advertising and free air time.

Promotion

- Developed outstandingly effective network of resources and support for Community Arts Festival, resulting in a lavish, "smash hit" fund raising event.
- Demonstrated gourmet food items and fine cookware in department stores and markets; educated public on its use; reported public reactions to manufacturers.
- Delivered product presentations to corporate employees during work hours, increasing customer base of our nearby store.

Customer Service • Needs Assessment

- Assessed clients' specific needs for catering services: handled initial inquiry, developed initial and final bids, visited site of affair, consulted with client to set desired menu and ambiance, proposed alternatives based on seasonal availability and unforeseen circumstances such as weather changes.
- Managed all aspects of large and small catered affairs, setting hosts totally at ease; oversaw food preparation; coordinated staffing, entertainment, setup, cleanup.

EMPLOYMENT HISTORY

1994–present	**PR/Promotions Dir.**	CYBELLE'S PIZZA RESTAURANTS, SF & East Bay
1992–present	**Sales Rep. & Partner**	CREATIVE MARKETING VENTURES, SF
1989–93	**Caterer/Sales Rep.**	LET THEM EAT CAKE, self-employed, Oregon & CA
1989–93	**Product Demonstrator**	Freelance for West Coast manufacturers
1989–92	**Fundraiser/PR Asst.**	KAREN WHITTMAN PR / ADV. CO., Portland, OR
1985–89	**PR/Promotions Dir.**	COMMUNITY CENTER RESTAURANT, Eugene, OR

EDUCATION

B.S., Education, Temple University, Philadelphia, PA
Graduate studies in Education and Administration, University of Oregon

CYNTHIA MICHAELS MAYER

2127 - 20th Street, Apt. 6
San Francisco, CA 94114
(415) 329-6667

Objective: Position in Marketing, Sales, or Client Services.

SUMMARY

- 6 years' professional business experience.
- Diplomatic and tactful in dealing with clients.
- Exceptionally patient and effective in training new staff.
- Thoroughly enjoy coordinating and managing projects.

RELEVANT EXPERIENCE

Marketing Support/Client Services

- Assembled customized marketing packets used in presentations to potential clients.
- Calculated and formatted data on firm's investment performance for marketing use.
- Oriented bank clients on services, management agreements, and account procedures.
- Successfully collected interest due to clients, previously lost through brokerage firm errors.
- Advised clients on tax questions, investment fees, bank statements.

Supervision and Training

- Supervised daily activities of staff, quickly shifting priorities as market fluctuated.
- Reviewed and approved reports to clients summarizing their account performance.
- Trained inexperienced Portfolio Assistants in complex administrative and operational procedures for handing clients' investment accounts, specifically how to:
  - Set up and maintain portfolio activity on the computer system;
  - Analyze the performance of an account;
  - Resolve stock delivery problems and bank statement discrepancies;
  - Analyze clients' accounts, identifying sources of ready cash or tax benefits;
  - Produce and process all legal documentation related to a client's portfolio.

Project Management

- Managed 3-month computer conversion project, communicating detailed specifications to programmers and supervising final transfer of data.
- Streamlined the process of gathering stock trading information and reduced costly errors, by designing better forms and procedures.
- Developed and coordinated complex stock buy-and-sell programs, involving accurate allocations of large sums of money, to achieve the Portfolio Manager's investment objectives.

WORK HISTORY

current	***Project Assistant*** *(part-time)* *(concurrent with relocation and career transition work)*	ALUMNAE RESOURCES Career Ctr., S.F.
1996	***Mgr., Portfolio Operations***	SNYDER CAPITAL MGMT. CO., NYC
1993–95	***Portfolio Assistant***	" " "
1991–92	***Customer Service Rep.***	METROPOLITAN TRUST CO., Cambridge, MA

EDUCATION

B.A., Urban Studies – Hobart and William Smith Colleges

Elizabeth mentions her current graduate studies because that's more directly relevant than her completed degree.

ELIZABETH JULIAN
1224 Seymour Avenue
Berkeley, CA 94709
(510) 345-6789

OBJECTIVE

Position as publicist with an arts organization.

SUMMARY

- Personable; work effectively with wide range of personalities.
- Experience in writing press releases and PSAs.
- Extensive contacts in the arts/entertainment field.
- Work well under pressure.
- Practical talent for seeing what needs to be done, and doing it.

RELEVANT EXPERIENCE

PUBLIC RELATIONS

- Composed press releases and public service announcements publicizing Bay Area musical benefits.
- Developed cooperative relationships with entertainment columnists from Bay Area news publications, resulting in successful coverage of musical events.
- Organized a calendar of advertising deadlines for several community organizations.

PROJECT ORGANIZING

- Compiled Directory of Bay Area Editors and publishers for use of literary agents.
- Organized educational workshop on publishing for writers and authors, involving:
  ...soliciting speakers ...participant mailing list ...space rental ...advertising.

WRITING

- Edited technical and literary book manuscripts, using word processor and WordStar program.
- Wrote and recorded copy for use as spot advertising on local radio station.
- Published poetry.

TECHNICAL SKILLS

- Filmed and edited 8mm film of a new, experimental music performance at Stanford Center for New Music Research (later transferred to video for use as instructional feedback).
- Videotaped and edited classes in multimedia performance, for instructional feedback.
- Designed and produced flyers to advertise benefits.

EMPLOYMENT HISTORY

1996–present	**TAP Admin. Assistant**	UC BERKELEY, Music Dept.
1990–95	**Editorial/PR Assistant**	BARRET, INC., LITERARY AGENTS, SF
1989–90	**Film maker (Video/8mm)**	Freelance – SF Bay Area
1987–89	**Benefits Organizer**	Freelance assignments for agencies – SF Bay Area
1985	**Aide to General Manager**	SAN FRANCISCO OPERA

EDUCATION

Graduate studies in expressive arts, in progress – JFK UNIVERSITY

B.S., Environmental Planning – UC Berkeley, 1987

Again, consistent layout goes a long way in bringing clarity to a complicated work history.

MARKETING & PROMOTION

FRAN MORGAN
1598 Waverley Avenue
Berkeley, CA 94709
(510) 523-3635

Objective: Position in sales/marketing, promotion or public relations, dealing with consumer services or high quality consumer products.

SUMMARY

- Personable and persuasive; able to build instant rapport.
- Aggressive, enthusiastic and energetic self-starter.
- Effective working both independently and as a team member.
- Successful experience in sales, marketing and promotion.

RELEVANT EXPERIENCE

Sales/Customer Relations

- Successfully sold expensive video-dating club memberships (minimum $1250) to men and women:
  - Interviewed, screened, and selected prospects;
  - Advised on affordable financial arrangements.
- Sold custom-made jewelry at the Whole Life Exposition, KPFA Crafts Fair, Women's Center Fair, etc., advising potential customers on appropriate colors, designs and gift purchases.
- Recruited top caliber applicants to graduate programs of UC Berkeley Business School, involving correspondence, phone interviews, school tours, faculty introductions and financial aid advice.

Promotion/PR

- Promoted Bike-a-thon for Cystic Fibrosis Foundation:
  - Assisted fund-raiser in contacting potential sponsors; distributed fliers and announcements;
  - Greeted participants at the event, collected contributions, distributed tee-shirts.
- Designed brochures, announcements and proposals to promote various activities:
  - Gathered and edited data for brochure describing research activities for state legislature;
  - Wrote text of announcement advertising programs at School of Business;
  - Edited grant proposals involving more than $500,000;
  - Assisted PR speech writer with idea development and editing for addresses by VP of Public Relations for the University.

Marketing/Display

- Designed and set up artistic displays of merchandise at annual arts and crafts fairs.
- Demonstrated boutique clothing and accessories at off-site locations (restaurants, student gatherings).
- Contracted with jewelry designer as sales rep to better department stores and boutiques.

– Continued –

EMPLOYMENT HISTORY

1987–present	***Administrative Asst.***	SCHOOL OF BUSINESS, UC Berkeley
1995 (part-time)	***Sales Representative***	MATCHMAKERS Video Dating, Oakland
1985–present	***Sales Rep*** (part-time)	KATYA JEWELRY, Oakland
1983–85	***Administrative Asst.***	VP PUBLIC RELATIONS Office, UC Berkeley
1982 (part-time)	***Model/Marketing Asst.***	RAGS LIMITED, Madison, WI
1981 (part-time)	***Social Worker Intern***	V.A. HOSPITAL, Madison, WI

EDUCATION & TRAINING

B.S., Psychology – University of Wisconsin, Madison, WI
Additional training: Graphic Design, Interior Design
Affiliation: National Association of Professional Saleswomen

GARY BRADLEY

1717 East Street
Concord, CA 94521
(510) 881-3637

Objective: Marketing or Sales Management.

SUMMARY OF QUALIFICATIONS

- 15 years' senior management experience in transportation and travel services.
- A born leader; effectively handled positions of major responsibility on a continuous path of professional advancement.
- Naturally creative; able to see the overall picture from initial concept through successful completion.
- Hands-on knowledge of virtually all positions within the industry.

MARKETING & SALES EXPERIENCE

Promotions

- Conceived and developed creative product promotions in travel and transportation:
  - Chartered Hornblower Yachts as site for highly successful presentation to 850 travel agents attending the Travel Age West convention;
  - Rented Playboy Club, attracting 1000 LA travel agents to sales presentation;
  - Scripted and delivered the presentation;
  - Led 250 travel agents in promotional excursion to Portugal, marketing golf packages.
- Directly implemented all phases of promotional programs:
  ...locale ...product ...visuals ...catering ...giveaways ...delivery of presentation.

Creative Advertising & PR

- Designed unique advertising with innovative placements, such as:
  - Aerial advertising of Hawaiian tour packages, at major sports events;
  - BART station billboards promoting London and Paris charters;
  - Multicolor ads in trade publications.
- Administered million-dollar advertising budget for Suntrips of California.
- Represented Suntrips, Global and TAP in presentations to both industry and media.

Sales

- Attained revenue quota of $56 million with Global exceeding revenue goal by 37%, where all other regions failed to achieve target.
- Set annual sales record for Suntrips in 1985, with revenues of $46 million.
- Established and administered $133 million sales budget for Global Airways.

MANAGEMENT & ADMINISTRATION

- Reorganized Suntrips of California, largest wholesale charter operator in the western United States, achieving No. 1 position in sales, with revenues of $50 million:
  - Designed and wrote new policy manuals and job descriptions for all departments;
  - Streamlined the delegation of responsibilities to existing managers;
  - Trained staff and managers in implementing more productive policies.
- Successfully directed both large and small field sales forces, consistently attaining revenue projections: –TAP-Air Portugal; –Global Airways; –Suntrips of California.

– Continued –

EMPLOYMENT HISTORY

1995–present	***Executive Vice President***	SUNTRIPS OF CALIFORNIA, San Jose
1995	***VP, Marketing & Sales***	SUNTRIPS OF CALIFORNIA, " "
1992–94	***VP, Marketing & Sales***	GLOBAL AIRWAYS/Western, Oakland
1990–92	***Dir., Agency Mktg. and Sales***	GLOBAL AIRWAYS " "
1987–90	***Area Sales Mgr., West. U.S.***	TAP Air Portugal, San Francisco
1979–86	***District Sales Manager***	TAP Air Portugal, Cleveland
1975–79	***Customer Services***	AMERICAN AIRLINES, Cleveland
1973–75	***Delivery & Dispatch***	UNITED PARCEL SERVICE, Cleveland

EDUCATION & TRAINING

Graduate, US Navy Dental Technology

Marketing & Management Training:
- Incentive Sales
- International Sales & Marketing
- Domestic Sales and Marketing
- Telemarketing
- Domestic & International Tariffs
- Management and Administration
- Passenger & Cargo Sales

GARY ROSEKRANS

3500 Derby Street
Berkeley, CA 94705
(510) 292-7860

Objective: Position as accounts executive with an ad agency or design firm.

SUMMARY OF QUALIFICATIONS

- Five years' successful experience in sales / marketing / advertising, with a special emphasis on point-of-purchase and display.
- Proven record in effectively handling major accounts.
- Conceptual talent and hands-on experience in sales driven projects.
- Able to elicit the trust and confidence of clients.
- Creative and resourceful in generating new ideas and solving problems.

PROFESSIONAL EXPERIENCE

Sales

- Developed successful sales strategy for Plastic Works' display products:
  - Researched and selected effective placement of advertising in national publications;
  - Coordinated all the design elements and directed image, content and copy of ads.
- Sold custom designed point-of-purchase programs and standard display products:
  - Researched target area and developed leads;
  - Made sales presentations to potential customers, outlining Plastic Works' design and manufacturing capabilities.
- Organized and participated in trade shows selling Plastic Works' products.

New Product Strategy

- Designed and produced innovative displays for the video software industry:
  - Researched the video industry and determined what products were needed, specifically:
    ...display fixtures for retail stores;
    ...point-of-purchase displays provided by major manufacturers;
  - Co-designed with industrial designer, compact and effective videotape display units;
  - Served on R&D team for manufacture of fixtures, and monitored production;
  - Marketed and sold the fixtures to stores and manufacturers.
- Currently researching the potential of developing displays for the video rental market in convenience stores and supermarkets; developed fixtures for that specific retail environment.

EMPLOYMENT HISTORY

1990–present	**Sales & Product Manager**	PLASTIC WORKS, Berkeley
1986–89	**Owner**	SIGNAGE CO. , Sign & Graphics, Oakland
1984–86	**Sign Fabricator/Installer**	THOMAS SWAN SIGN CO., San Francisco

EDUCATION

SAN FRANCISCO ART INSTITUTE, Photography / Design

Additional Studies – UC Berkeley Extension:
• Marketing & New Product Development • Advertising Strategy

GERALD DAVIS

7600 Miracle Road ◆ Napa, CA 94558 ◆ (707) 899-6000

Objective: **Position in Sales & Marketing,** focusing on management, supervision, sales and product development.

SUMMARY

- 12 years' experience as Vice President of Sales & Marketing.
- Successful in generating new business and increasing sales volume.
- Effective in persuading others through my enthusiasm.
- Highly reliable; proven ability to set and meet goals.

PROFESSIONAL EXPERIENCE

1994–present **VP Sales & Marketing** CALNAP TANNING CO. – Napa, CA

- Developed new markets for leather, expanding from shoes and handbags to a wide range of other related products, significantly increasing sales.
- Researched clothing and shoe market in field trips to Europe; designed and created wide range of new colors to coordinate with current trends.
- Introduced new weights of leathers which increased our company's share of the market in heavy-weight leather products.
- Increased leather sales from $7 million to $18 million during first 6 years, closing many difficult sales by effectively overcoming objections.
- Introduced new and existing lines of leather all over the U.S., making fashion presentations to marketing directors of major manufacturers.

1990–93 **Product Manager** PHILIP A. HUNT CHEMICAL CORP. – SF

- Developed and supervised five sales agencies throughout the country:
  – traveled to each territory regularly – trained sales reps – visited major accounts.
- Coordinated and supervised production, overseeing quality control and scheduling.
- Monitored overall profitability, accurately projecting manufacturing costs and product pricing.

1987–90 **Technical Sales Rep** EASTMAN KODAK CO. – Los Angeles

- Opened up several major profitable new accounts on the East Coast:
  – Introduced our products to companies previously unaware of us;
  – Overcame distance barrier by offering persuasive advantages such as:
  ...modern facilities ...high quality service ...personalized attention.

EDUCATION & TRAINING

B.S., Business Administration – UNIVERSITY OF CALIFORNIA, Sacramento
Graduate, Dale Carnegie Sales Course
Industrial Psychology workshop

Betsy's work history illustrates both the variety of her roles AND her increasing level of responsibility.

BETSY EMORY

32 Center Street
Oneonta, NY 13820
(607) 402-7442

Objective: Position as Project Director or Public Relations Officer with a public service agency.

SUMMARY

- Familiar with the City of Oneonta; active resident for 23 years; take personal pride in representing my community.
- Experienced manager; effective in delegating and developing staff skills.
- Successful in generating good will and restoring confidence.
- Sincerely enjoy the challenge of working with people.
- Respond effectively and creatively to change.
- Able to focus on specific tasks, keeping overall project goals in mind.

RELEVANT PROFESSIONAL EXPERIENCE

MANAGING/DIRECTING

- Managed a sales/marketing staff of five: two account managers covering eastern and western national regions; two part-time telemarketing employees; art director.
- Oversaw and evaluated the effectiveness of national distribution network.
- Monitored sales/marketing budget for base of $7 million product sales.

PUBLIC RELATIONS

- Represented company on extended sales trips: met with clients, promoted good will with retailers, presented new products.
- Organized and planned convention displays and strategy.
- Designed and developed a telemarketing program for Briarpatch Natural Foods:
  – Researched needs of the natural foods marketplace;
  – Placed new products in areas not previously serviced.

PROJECT DEVELOPMENT

- Conceived promotional graphics ideas: ads, posters, point-of-purchase material, newsletters; oversaw completion of the projects in coordination with company art director.
- Developed and delegated direct retail monthly mailing program: promotional calendars, promotional fliers, and product samples.

EMPLOYMENT HISTORY

1994–present	*National Sales Manager*	BRIARPATCH NATURAL FOODS, Oneonta, NY
1993–94	*Northeast Account Manager*	" " "
1990–93	*Customer Service Coordinator*	" " "
1989–90	*Asst. Accounting Manager*	" " "
1987–89	*Accounts Payable Clerk*	" " "
1986–87	*Computer Operator/Receptionist*	" " "

EDUCATION & TRAINING

B.A., Political Science; Art – STATE UNIVERSITY of NEW YORK, Albany
Numerous marketing seminars, through American Marketing Association

MARKETING & PROMOTION

JEAN BOGART

#7 Sassoon Drive
San Francisco, CA 94132
(415) 225-8735

Objective: Position as Client Services Rep. with Metropolitan Medical Center.

SUMMARY

- Twenty years' successful experience in direct sales.
- Proven effective in public relations targeted to the medical community.
- Superior knowledge in use of chemical dependency treatment programs.
- Strong public presentation skills.
- Able to start up a program from scratch.

PROFESSIONAL EXPERIENCE

Sales & Client Base Development

- Built and maintained client bases for four service-oriented businesses—
  group travel, stock brokerage, funeral services, training and development:
  – Identified the market via a demographic survey, and developed needs analysis;
  – Designed a sales presentation to fit identified needs of clients;
  – Maintained follow-up program involving newsletters, seminars, personal contact.
- Developed sales from zero and one location, to million-dollar annual sales and three locations, in 13 years.
- Opened a brokerage office and built it to $10,000/month revenue in 6 months.

Presentation & Health Education

- Facilitated week-long courses on Substance Abuse Prevention for US Navy.
- Presented employee workshops: Listening, Managing Assertively, Self Awareness.
- Spoke before numerous community groups on family mental health issues.
- Designed and delivered marketing presentations on financial planning to business and service groups.

Marketing & Public Relations

- Designed highly successful marketing program for municipal Employee Services.
- Collaborated in the design and evaluation of a training program for UC Berkeley certificate program in Training and Human Resource Development.
- Defined the desired public image goals of hospital, church, and educator groups, and designed effective marketing and sales presentations consistent with their goals.

EMPLOYMENT HISTORY

current	**Employee Services Intern**	CITY OF OAKLAND, CA
1993–96	**Independent Consultant**	Training & Human Resource Development, Bay Area
1990–92	**Stock Broker**	E.L. TRUMBUL & SONS, member NY Stock Exchange
1988–90	**Owner/Operator**	PRIMO TOURS, Camarillo, CA
1973–90	**VP, Personnel/Sales**	BAY FUNERAL SERVICES, Oxnard, SF, San Diego

EDUCATION

B.A., Psychology – UC Santa Barbara
Certificate, Training & Human Resource Development – UC Berkeley
Certificate, Alcohol and Chemical Dependency – JFK University, Orinda

MARKETING & PROMOTION

JOANNE SIMPSON

1609 Walnut Street, Apt. 3
San Francisco, CA 94123
(415) 377-2882

Objective: Entry level position in a market research firm, leading to account management.

SUMMARY

- 8 year background in retailing and wholesaling.
- Degree in Business with concentration in Marketing.
- High level of enthusiasm and commitment to a marketing career.
- Strong leadership qualities; able to take charge and get things done.
- Broad perspective of people and markets, based on extensive travel.

RELEVANT EXPERIENCE

Sales

- Sold Clinique cosmetic products in three major retail stores:
  – Demonstrated the product to individual customers, advising on colors and use of skin care products;
  – Displayed merchandise in cases and throughout the department.
- Represented Clinique during promotional events at Bay Area retail stores, advising customers and doing product demonstrations.
- Prepared monthly sales reports, wrote purchase orders, and maintained stock control book.

Management

- Developed an understanding of group dynamics, individual motivation and interpersonal communication skills, in classes on Management and Human Behavior in Organizational Settings.
- Managed Clinique counter at Bullock's and The Crescent, involving supervision of two sales employees, staff motivation, and achieving sales goals.

Promotion and Advertising

- Developed a media plan as a component of an advertising campaign for a food product (class assignment), addressing three explicit marketing objectives:
  ...Obtain greater distribution ...Increase market share ...Increase sales by 5%.

EMPLOYMENT HISTORY

1995–present	**Retail Sales**	MACY'S Clinique Counter, San Francisco
1994	**Promotional Asst.**	CLINIQUE COSMETICS, Bay Area
1993–94	**Asst. to Office Mgr.**	MARKETING VP of Bremworth Carpets, SF
1989–93	**Counter Manager**	BULLOCK'S Clinique Counter, SF
1988–89	**Counter Manager**	CRESCENT Dept. Store, Clinique Ctr., Spokane, WA
1987–88	**Flight Attendant**	NORTHWEST ORIENT, Minneapolis, MN

EDUCATION

B.S., Business – concentration in Marketing, San Francisco State University

– References available upon request –

Travel and full-time student status fill in the work history, leaving no gaps.

JOANNE FINE

220 Georgetown Street • Albany, CA 94706 • (510) 529-0620

Objective: Professional affiliation with Holbrook Associates assisting in international marketing expansion.

PROFILE

- Successful generalist, with specialties focusing on:
  ...Development of third world countries;
  ...Agricultural economics; Natural resources management;
  ...Social, political and economic factors of nutrition.
- Fluency in French and Spanish gained from living abroad.
- Special talent for relating well with people of diverse interests.
- Creative and resourceful in generating new ideas and solving problems.
- Degree in Conservation and Resource Studies.

PROFESSIONAL EXPERIENCE

Marketing / Public Speaking

- Successfully marketed and expanded EIP/Northern California program, exceeding project goals by 50% and increasing project income by 75%.
- Competed successfully, via RFPs, for projects with state and local governments.
- Spoke before many professional groups, introducing the organizations' programs; appeared on many panel discussions on careers in the environmental field.

Project Development

- Assessed needs of potential EIP project sponsors, determining specific personnel requirements, cost and length of project, work objectives, and expected results.
- Recruited and screened hundreds of potential short-term employees, matching their skills and potential for professional growth, with sponsors' identified needs.
- Followed up with on-site visits to ascertain that both sponsor and employees were satisfied with the implementation of the project.
- Originated special "Minority Urban Environmental Program" to encourage minority professionals to enter environmental occupations—accepted for funding by the San Francisco Foundation.

EMPLOYMENT HISTORY

1994–present	**Regional Director**	EIP/Northern California (Environmental Intern Progr), SF
1993	Travel	Europe
1990–92	**Full-time student**	UC Berkeley, Conservation and Resource Studies
1978–90	**Admin. Assistant**	UC Berkeley – Office of the Dean of Letters/Sciences
1976–78	**Office Manager**	Nat'l Democratic Committeeman New Orleans, LA

EDUCATION

B.S., Conservation and Resource Studies, UC Berkeley – Phi Beta Kappa, 1992

MARKETING & PROMOTION

ELIZABETH LEONARD

340 California Street • Sacramento, CA 95818 • (916) 881-7213

OBJECTIVE **Position in public relations, public affairs or promotions.**

SUMMARY

- Over two years' successful experience in public relations.
- Special talent for persuasion and problem solving.
- Relate easily with all kinds of people, as company representative.
- Skilled in writing PSAs and promotional material.
- Creative, energetic, positive and hard working.

SKILLS AND EXPERIENCE

Public Relations/Problem Solving

- Successfully handled PR problems for cable TV company, gaining the cooperation of previously resistant homeowners, for installations on their property:
  - Established friendly communication and identified homeowners' specific objections;
  - Negotiated creative solutions acceptable to both our company and homeowners.
- Resolved restaurant's PR problem involving customer injury, demonstrating genuine concern for the customer, taking responsibility for medical costs, and successfully retaining the good will and business of the customer.

Promotion

- Promoted campus entertainment events:
  - Wrote PSAs and ads; – Implemented creative promotional ideas;
  - Designed and distributed fliers.
- Sold program advertising space for a fund-raising musical event, raising money for Stanford Children's Home.
- Promoted special seasonal offerings for a gourmet vegetarian restaurant:
  - Proposed new entrees; – Designed menu;.
  - Designed and distributed discount coupons.
- Currently developing a 60-second TV spot to raise funds for a local charity.

Project Management/Organization

- Coordinated programming and scheduling for a KGNR live radio talk show:
  - Contacted public figures and ordinary citizens to set up specific schedule;
  - Wrote up biographical material and proposed questions, for radio anchorman;
  - Followed up to confirm appointments just prior to show time.
- Managed Mum's in Sacramento, an 80-seat restaurant:
  - Hired, supervised and scheduled employees;
  - Monitored customer satisfaction.

WORK HISTORY

1995–present	***Construction Coordinator***	SACRAMENTO CABLE TV, Sacramento
1993–95	***Restaurant Manager***	MUM'S RESTAURANT, Sacramento
	Concurrently with:	
1993–94	***Producer Intern***	KGNR RADIO, Sacramento
1994 spring	***Public Relations Intern***	STANFORD CHILDREN'S HOME, Sacramento
1990–94	***Student***	CAL STATE UNIVERSITY, Sacramento

EDUCATION B.A., Communication Studies – CAL STATE UNIVERSITY, Sacramento

LESLIE ROSE BOWMAN

990 Mystic Mountain Dr. • Mill Valley, CA 94941
(415) 278-5772

Objective: Production Coordinator for Fashion Special Events and Public Relations.

PROFILE

- Outstanding stylist with a passion for art and clothing.
- Effective as both project director and in cooperative teamwork.
- Lifelong exposure to couture fashion and retailing.
- Portfolio of current, forward, innovative fashion photography.
- World traveled; worked with international photographers.

RELEVANT EXPERIENCE

Special Events Coordinating & PR

- Coordinated a successful, major fashion show for SOFTWEAR art-to-wear gallery: conceived the idea of bringing together two award-winning weavers with a leading San Francisco fashion designer, in a successful collaborative showing.
- Implemented a highly effective showing of a relatively unknown but talented weaver, greatly enhancing her visibility and professional opportunities.

Management & Production

- Ran a retail clothing store specializing in artistic one-of-a-kind items:
  - Bought and sold all the clothing and accessories; set up all the displays;
  - Established a better financial arrangement with artists which compensated them immediately, benefiting both parties and improving production.

Fashion Styling

- Produced three ads for Diet Center that appeared in Vogue, Harper's Bazaar, Woman's Day, Working Woman, McCalls and Glamour magazines.
- Styled wardrobe for Levi Strauss video, publicizing their shirts.
- Advised private clients on wardrobe for evolving career and fashion images.

Artistic Creativity

- Managed production of 6 album cover photographs: hired photographer; engaged makeup & hair stylists; found location; assembled props; styled the cover and back.
- Originated a unique pants design adapted from traditional Moroccan pattern, now in widespread use.
- Designed stage costumes for famous entertainers which helped create appropriate public image.

WORK HISTORY

1994–present	**Wardrobe/Fashion Consultant**	Freelance, San Francisco
1994	**Manager**	SOFTWEAR & ARTWORKS, SF (Art-to-wear clothing, accessories, artifacts)
1993	**Saleswoman**	KEBAYA Co., imported clothing, Sausalito
1985–92	**Tours Coordinator**	Part-time, while raising children
1978–79	**Chief Docent/ Asst. Curator**	ISABELLA S. GARDNER MUSEUM, Boston

EDUCATION

Theater lighting design, stage makeup, stage costume design – College of Marin
Painting, sculpture, and photography classes – San Francisco Art Institute
Painting, sculpture, art history – Philadelphia College of Art

LINDA MacKINNON
7907 Broderick Street
San Francisco, CA 94115
(415) 994-4521

Objective: **Client services representative**, promoting the programs/services of a hospital, clinic or health services association.

SUMMARY

- 5 years' experience in marketing health services and products.
- Outstanding record of success in outside sales.
- Persuasive and knowledgeable in health services presentation.
- Communicate well with business professionals, easily establishing rapport and gaining client confidence.

SALES / MARKETING EXPERIENCE

Product Presentation/Demonstration

- Made presentations to administrators of HMOs, IPAs and medical laboratories, on the advantages of Keller's outpatient and lab management programs.
- Explained wide range of test methodologies to doctors and lab technicians.

Direct Sales

- Successfully persuaded major SF employers to sponsor blood drives:
  - Convinced them of the community service value of participation;
  - Explained the advantage to employees of accumulating blood bank credit;
  - Significantly increased employee participation over previous drives.
- Made cold calls to physicians and laboratory directors of hospitals and clinics to market reference laboratory services.

Client Services

- Serviced client accounts as sales rep for Sebring Sales, Inc.:
  - Made follow-up visits, inviting feedback on satisfaction with services;
  - Responded quickly to resolve clients' problems;
  - Maintained rapport by showing personal interest in clients and their business.
- Assisted company sponsors in coordinating logistics of on-site blood drives:
  - Prepared publicity – Supervised staffing – Gave educational talks to employees.

EMPLOYMENT HISTORY

1995–present	***Sales Rep***	KELLER BIOMEDICAL LABS., Sacramento
1994	***Field Recruiter***	IRWIN MEMORIAL BLOOD BANK, SF
1991–93	***Sales Rep***	SEBRING SALES (manufacturing reps.), SF
1989–90	***Planner/Print Buyer***	D'ARCY MacMANUS ad agency, SF
1988–89	***Executive Secretary***	MJB COMPANY, SF

EDUCATION

A.A., Psychology, Santa Barbara City College
Biology minor, University of California, Santa Barbara

SANDRA CERRITO

P.O. Box 1213A • San Francisco, CA 94126 • (415) 808-5772

Objective: Marketing Manager in the medical products industry.

MARKETING EXPERIENCE

Marketing Management

- As Product Manager, participated in strategy sessions to identify NL Industries' growth product lines:
  - Conducted market segmentation studies;
  - Developed analysis of our resources and strengths compared with competitors;
  - Recommended to management whether to enter this market area;
  - Participated in Manager's Roundtable on effective marketing to key accounts.
- Analyzed field experience with specialty industrial drilling chemical for NL Industries, an oil field services company; compared laboratory data; made recommendations on product development and marketing; outlined sales promotion program.
- Chaired regional conference on field marketing strategy as Marketing Consultant for ECKART Company, resulting in recommendations to corporate officers.
- Wrote sales training bulletins on better customer relations, effective communication, and sales strategy, which were distributed to Pfizer's western regional area.
- Researched and wrote an overview of Pfizer hiring practices and made recommendations to VP of Personnel, on improving company record of hiring women.

Professional Sales

- Won top award at Pfizer Company for sales volume increase in metropolitan Chicago area.
- Re-established professional relationship and trust with 50-60 neglected pharmaceutical accounts, tripling Pfizer's sales volume and increasing profits in my territory.
- Consistently achieved top ranking in Pfizer district sales competition; trained new salespeople in the field.

EMPLOYMENT HISTORY

1994–present	**Staff Assistant**	AD VENTURES INC. (venture capital company), SF
1993	**Marketing Consultant**	Independent contractor, ECKART CO., Houston, TX, implementing sales campaign for executive benefits
1991–92	**Product Manager**	NL INDUSTRIES (oil field services) – Houston, TX
"	**Sr. Business Analyst**	" " "
1990	**Regulatory Analyst**	AMERICAN HOSPITAL ASSN. – Chicago, IL
1982–89	**Senior Sales Representative**	PFIZER CO. – Chicago, IL
1980–81	**Pharmaceutical Chemist**	ELI LILLY & CO. – Indianapolis, IN

EDUCATION

B.S., Biology – Washington University, St. Louis, MO, 1979

Seminars & Conferences:

- New Product Development, and Direct Mail Marketing – UC Berkeley, 1994
- Product Management Conference – American Marketing Association, 1992
- Strategic Market Planning – Braxton Associates, 1991
- Sales Forecasting – University of Houston, 1989

Sales Resumes

Amy Kurle	Sales and customer service	181
Deborah Richardson	Corporate sales rep, hotel	182
Denise Walters	Sales rep, Kodak	183
Donna Cole	Sales/medical product	184
Ellen Metcalfe	Sales rep/manufacturer's rep	185
Hollis Ann Pope	Buyer/sales, merchandise	186
Jerry Wilcox	Outside sales rep	187
Jerry Parkhurst	Merchandising display	188
Tricia Baker	Sales management, consumer services	189
Judy Rogers	Sales/marketing	190
Linda Mowry	Sales rep/account executive	191
Mark Fleetwood	Sales rep, electronics	192
Melinda Sailor	Jewelry sales/customer service	193
Noreen MacLaughlin	Sales	194
Pamela Swiss	Outside sales rep, fashion	195
Sherrie Valencia	Sales/marketing	196
Vreny Zurich	Small store manager, shoes	197
Eleanor Kennedy	Sales rep	198

AMY PARKER KURLE

19000 Sixth Ave. NW
Bremerton, WA 98888
(206) 777-7544

OBJECTIVE

Sales / Customer Service Manager, Auto Body Specialists

HIGHLIGHTS

- 15 years experience in the auto parts and service industry.
- Professional attitude toward customer satisfaction, resulting in an excellent reputation with customers.
- Ability to balance books and handle finances in a responsible manner.
- Purchasing experience and expert knowledge of automotive parts.

RELEVANT EXPERIENCE

1990–present **J & D DISTRIBUTORS**, Fremont CA (import parts wholesaler)
Inside & Outside Sales

Part of a two-person sales team with over a million dollars in sales per year, topping two million the last year.

- Successfully handled busy phones daily, servicing customers while meeting shipping and delivery deadlines.
- Maintained acceptable profit margins without alienating customers, through superior customer service.

1988-89 **BAYSIDE AUTO PARTS**, Piedmont CA (import parts wholesaler)
Domestic Purchasing and Inside & Outside Sales

Advanced from order-taker to outside sales person and eventually to domestic purchasing manager.

- Monitored inventory, requested and evaluated price quotes, prepared and placed purchase orders.
- Oversaw receiving procedures and maintained quality control of domestically purchased products.

1982-83 **F & W ENGINES**, Berkeley CA (VW & Japanese engine rebuilder)
Parts, Service Writing, Bookkeeping, Mechanics

Performed minor mechanical repairs and set-ups for machine work, as well as managing all phases of daily office operations, including:

- Service writing, estimating, and scheduling of work.
- Bookkeeping, banking, and accounts payable.
- Ordering, receiving, and keeping inventory of parts and supplies.

1981 **IMPORT CAR CENTER**, El Cerrito CA (retail import auto parts)
Driver, Counter Sales

Worked with retail customers as counter person.

- Received and stocked parts. Made deliveries.

DEBORAH RICHARDSON

3445 Mariposa Avenue • San Mateo, CA 94403 • (415) 446-9133

Objective: Corporate sales representative position in hotel sales/catering.

SUMMARY

- 8 years' successful experience in both inside and outside hotel sales; special talent for recapturing lost accounts.
- Professional and self-confident in handling corporate accounts.
- Working knowledge of major corporations in the S.F. Bay Area.
- Exceptional success in establishing rapport with clients.
- Able to work independently.

SALES ACCOMPLISHMENTS

1995–present **Account Representative** CALIFORNIA FURNITURE RENTAL, Foster City, CA

- Organized and implemented a highly effective 3-week, door-to-door "sales blitz" involving 10 sales representatives:
  – Researched 400 accounts to update reps on background information and likely problems;
  – Conducted strategy meetings; assigned territories and objectives for each rep;
  – Reviewed all incoming reports and submitted final analysis to management.
- Produced and published a promotional newsletter for corporate clients:
  – Wrote articles introducing their new rep and informing them of upcoming special events and discounts; – took photographs; – designed layouts.

1992–95 **Account Executive** DUNFEY HOTEL, San Mateo, CA

- Successfully regained Dunfey Hotel's largest account, as first sales assignment:
  – Opened a good line of communication by persistent contact;
  – Determined the cause of the problem;
  – Assured that the hotel knew and could meet the client's needs.
- Consistently exceeded sales and profit goals for corporate bookings.
- Created a new sales incentive program that effectively kept existing accounts, featuring attractive gift certificates for trips, meals, and hotel accommodations.
- Delivered presentations to groups of corporate executives, outlining hotel services available and conducting question-and-answer sessions.

1991–92 **Food/Beverage Restaurant Mgr.** DUNFEY HOTEL, San Mateo, CA

- Planned and directed all aspects of a party for 600 clients: created a theme; designed costumes and invitations; developed mailing list and menu, acted as emcee and hostess.

Prior Work History

1988–91 Executive Secretary Rodeway International, Omaha, NE
1985–88 Legal Secretary H. Neuhaus, Esq., Omaha, NE (acc't. recovery)

EDUCATION

Liberal Arts, IOWA STATE UNIVERSITY – Ames, IA

DENISE WALTERS
2330 - 43rd Avenue
San Francisco, CA 94121
(415) 233-2054

OBJECTIVE

Position as a Sales Representative with the Eastman Kodak Company Graphics Imaging Division.

SUMMARY

- Education and talent in the field of Graphic Arts.
- Resourceful and committed; can be counted on to get the job done.
- Self-motivated and well organized; enjoy the challenge of outside sales.
- Effective in delivering presentations that generate new business.
- Sharp, poised, able to convey a warm yet professional image.

SALES EXPERIENCE

DIRECT SALES / ACCOUNT MANAGEMENT

- Called on established key accounts (corporate and insurance):
  - Identified clients' needs and problems, assuring them of a personal representative they could count on;
  - Demonstrated a personal interest in clients, taking them to lunch and remembering personal details;
  - Resolved service problems, billing problems, and misunderstandings.
- Delivered effective sales presentations to business groups and individuals:
  - Introduced and promoted our product/service to leading insurance co's.;
  - Identified appropriate corporate officers and made introductory calls to establish new accounts.
- Increased account base by 50% at two locations, through assertive salesmanship and consistent follow-up.

PLANNING / REPORTING

- Developed monthly sales plans: set goals; identified account maintenance needed; targeted special problems requiring attention; set up schedule of appointments.
- Forecast sales goals by dollar amount as well as by specific referral sources.
- Maintained detailed daily sales logs and referral logs.
- Wrote extensive monthly sales reports, including calls made, problems identified, and plans for the coming month.
- Conducted detailed market analyses and monthly surveys of competition throughout the Bay Area, and submitted reports to regional manager.

EMPLOYMENT HISTORY

1995–present	**Sales Representative**	SUPERIOR RENT-A-CAR, Oakland
1993–94	**Sales Representative**	PERSONNEL POOL temp service, San Mateo
1992	**Sales/Service Manager**	CERTIFIED TEMPORARY PERSONNEL, San Bruno
1989–91	**Sales Representative**	GRANTREE FURNITURE RENTAL, San Mateo
1987–88	**Leasing Agent**	LINCOLN PROPERTY CO., San Bruno
1986	**Salesperson**	TOPS & TROUSERS, San Francisco

EDUCATION

ACADEMY OF ART COLLEGE, San Francisco
Major: Graphic Design, 1985-89

See Donna's cover letter on page 271.

DONNA COLE

1776 - 12th Avenue • San Francisco, CA 94118
(415) 212-6822

Objective: Sales position with a company marketing medical products.

SUMMARY

- Experience marketing to professionals in the medical field.
- Strong presentation skills; professional appearance and manner.
- Versatile and adaptable; welcome the challenge of solving problems.
- Proven record in building and maintaining a client base.

PROFESSIONAL EXPERIENCE & SKILLS

Sales and Promotion

- Made cold calls and field visits to medical wholesalers, significantly increasing accounts.
- Increased sales through effective demonstrations of fabricating techniques.
- Visited and serviced existing accounts to assure continued product sales.
- Advised clients on options available to meet a wide variety of patient needs.
- Acted as technical liaison between doctors and prosthetists.
- Followed up by phone on dealers' requests for information and samples.

Project Management

- Successfully assumed emergency interim management of production department following sudden loss of staff; maintained production schedules through prioritizing and rescheduling.
- Researched competition's design, fabrication materials and marketing techniques by visiting American Cancer Society facility and interviewing patients and staff.
- Maintained good customer relations by assisting clients in identifying technical and administrative problems, analyzing patient needs and developing solutions.

Business

- Set up a small business bookkeeping system:
  - – Acquired necessary licenses; – Handled accounts payable and accounts receivable;
  - – Consulted with accountants; – Processed payroll and disbursed commissions.
- Entered and retrieved computerized data.
- Recruited and supervised staff of 15 clerical and sales personnel.

SALES

EMPLOYMENT HISTORY

1992–present	***Prosthetic Technician/Asst. Mgr.***	GRIFFHEIMERS, San Francisco
1989–1992	***Prosthetic Technician/Sales Mastectomy Products***	HOSMER-DORRANCE CORP., Campbell " "
1989	***Bookkeeper***	ALEX'S PORSCHE HOUSE, Campbell
1987–1988	***Office Manager***	SAVE ON SOLAR, Inc., San Jose
1986–1987	***Bookkeeper***	PAUL COLE WRECKING, San Jose

EDUCATION

Small Business Management classes – Contra Costa College & San Jose City College
Registered Technician* – American Board of Prosthetics and Orthotics
*one of only six women currently registered in the country

ELLEN METCALFE

1912 - 20th Avenue ■ San Francisco, CA 94118
(415) 212-6755 home ■ (415) 494-0091 work

Objective: Position as sales representative or manufacturer's representative.

SUMMARY

- 8 years' successful experience in retail and business sales.
- Ability to establish instant credibility.
- Confident, professional business communicator.
- Special talent for identifying clients' needs and presenting effective solutions.

SALES & MARKETING ACCOMPLISHMENTS

1995–present **Communications Mgr.** HOME DESIGN CONSULTANTS, S.F.

- Made on-site presentations at residential and business locations to evaluate site requirements and provide customized space planning services.
- Addressed large and small groups at major trade shows and at product seminars.

1993–95 **Owner/Manager** GREGG WALL COVERINGS, Boise, ID

- Discovered a void in the home and office furnishings market, and created this successful business solving storage problems created by new technology.
- Conducted high-energy cold calling campaign, averaging 10-20 calls per day and successfully opening up new sales territory.

1992–93 **Branch Manager/Sales** BAYLIN Personnel Service, S.F.

- Negotiated acceptable solutions for several serious problem accounts with a long history of nonpayment and potential litigation against the company:
  – Restored relationships by reestablishing supportive contact;
  – Solved 20% of the problems by upgrading and selling additional equipment;
  – Renegotiated the terms of payment, effectively reducing receivables by 40%.

1990–92 **Sales Representative** CROWN COMPUTERS, San Mateo

- Researched and developed markets to build a client base:
  – Built a network of other professionals within my field and shared sources;
- Demonstrated features and benefits of computer products to corporate decision makers, and successfully negotiated favorable contracts.
  – Stayed abreast of industry developments through newspapers and trade journals.

1988–90 **Sales Representative** G.L.W. OFFICE SYSTEMS, Palo Alto

- Established highly effective relationships with potential clients at all levels, from support staff through management, employing a natural conversational style:
  – Created immediate rapport by establishing a commonality of interest;
  – Probed for an overview of business operations to assess client needs;
  – Skillfully closed sales by gaining agreement on the benefits of the product.

EDUCATION

B.A., English, PENNSYLVANIA STATE UNIVERSITY

Holly includes her middle name to head off confusion over whether "Hollis" is a man or a woman.

HOLLIS ANN POPE
6778 Grand Avenue
Oakland, CA 94609
(510) 699-2121

Objective: Position as associate in merchandise sales or buying, with special emphasis on client services, needs assessment and negotiations.

PROFILE

- Successful, varied experience in buying and negotiating.
- Ambitious, adventurous, goal- and profit-oriented.
- Outstanding people skills: sensitive in assessing needs.
- Readily inspire the trust and confidence of clients.
- Committed to professional excellence.

RELEVANT EXPERIENCE

Buying

- Researched and located potential overseas wholesale suppliers of high-quality leather clothing:
  – Reviewed wide range of designs, selected best combination of style, quality, price, and on-going availability;
  – Set up trade and banking operations between Turkish company and my business.

Sales

- Sold specialty clothing to retail store buyers, initiating contacts and making direct sales.
- Developed substantial and profitable repeat business, delivering high-quality service to restaurant customers:
  – Completed course in wines; offered expert advice on selections;
  – Observed customers' needs/wants, and personalized service accordingly.

Negotiating

- Bargained assertively and effectively with foreign merchants.
- Negotiated diplomatically with customers in an exclusive, limited-seating restaurant, maintaining both capacity seating and customer satisfaction.

Counseling/Needs Assessment

- Completed training in counseling skills at UC Berkeley Dept. of Education, and developed expertise in helping individuals communicate needs and problems.
- Helped students assess problem situations and develop practical plans for achieving positive change.

EMPLOYMENT HISTORY

1995–present	**Assistant Manager & Maitre d'**	BLUE DANUBE CAFE, Oakland
1993–94	**Waitress**	THACKERAY'S RESTAURANT, Oakland
1992–93	**Sales Assistant**	MYRA TREVOR IMPORTS, Berkeley
1989–92	**Owner/Manager**	POPE & LEWIS Importers, Berkeley
1985–89	**Full-time Student**	UC BERKELEY

EDUCATION

B.A., Psychology – UNIVERSITY of CALIFORNIA, BERKELEY,
Honors: Phi Beta Kappa

Jerry had an uncomfortable "image" problem: people's surprised reaction to "truck-driver" on his resume. We sacrificed consistency in form and changed the job-title to "transportation"

JERRY WILCOX

331 Montecito Avenue
Pleasanton, CA 94566
(510) 943-0104

Objective: Position as outside sales representative with a manufacturer or distributor.

SUMMARY OF QUALIFICATIONS

- 7 years' highly successful experience in direct outside sales.
- One of the top salesmen nationwide with Coca Cola.
- No. 1 in sales with Smith Corona Marchant's western U.S. region.
- Highly motivated; an achiever who sets and reaches his goals.
- Proven skills in problem solving and customer relations.

SALES EXPERIENCE

Direct Sales & Account Management

- Trained new sales reps for developing new and existing territories, at both Smith Corona and Coca Cola.
- Achieved status of top salesman nationwide with Coca Cola:
  - Sold more Coca Cola coolers than any prior salesman;
  - Personally called on every customer on a weekly basis, maintaining good public relations and reviewing customers' needs;
  - Designed customized exterior and interior signs featuring customers' products.

Problem Solving & Customer Service

- Increased SCM's territory sales by 120%, identifying clients' specific needs and designing cost effective alternatives:
  - Demonstrated opportunities to save time and money in office billing procedures;
  - Completely revised and automated clients' billing systems;
  - Sold new SCM copying equipment required for the newly installed system.

Community Service

- Served actively in a wide range of community service organizations and commissions.
- Ran for City Council in 1992 and 1994.
- Appointed as Commissioner for prestigious Alameda Co. commission, responsible for:
  - Allocating $800 million in retirement fund investments;
  - Presiding at monthly hearings and making final judgments on disability cases.

EMPLOYMENT HISTORY

1982–present	***Transportation***	WOODLAKE MANUFACTURING, Oakland
1978–81	***District Sales Rep***	SMITH CORONA MARCHANT (SCM), Oakland
1973–78	***District Salesman***	COCA COLA CO., Hayward

EDUCATION & TRAINING

Sales training course with Smith Corona

SALES

JERRY D. PARKHURST

803 Azalea Drive
El Cerrito, CA 94530
(510) 296-7566

Objective: Position in merchandising display, with a manufacturer, distributor or advertising agency.

SUMMARY OF QUALIFICATIONS

- Three years' experience in setting up creative, effective merchandising displays.
- Self-motivated, honest and dependable.
- Working familiarity with the Bay Area.
- Successful in maintaining rapport with retailers.
- Well groomed and professional in manner.

RELATED EXPERIENCE

Display

- Set up effective retail displays of beverages in supermarkets, liquor barns, liquor stores, and package stores on military bases.
- Inventoried and reordered display materials, and maintained warehouse, for Glenmore Distilleries.

Customer Relations

- Developed cooperative working relationships with retail owners and managers:
  - Introduced myself as merchandising representative;
  - Advised managers of available promotional theme;
  - Pointed out the benefits of the promotions, such as increased sales, and secured approval for displays.

Servicing Existing Accounts

- Assured that products were priced, positioned in proper location, and that adequate stock was on hand.
- Maintained accurate route sheet with return dates, and refurbished displays on a regular two-week schedule.
- Serviced existing accounts in Contra Costa, Alameda, Solano, and Napa Counties.

WORK HISTORY

1995–present	**Apprentice Mechanic**	SKYTREADS, aircraft wheels and brakes, SF
1993–95	**Merchandiser**	GLENMORE DISTILLERIES, Richmond
1992–93	**Merchandiser**	BEVERAGE DISPLAY, San Mateo
1989–92	**Mechanical Assembler**	SYNMED, INC., optometry machine mfg., Berkeley
1988	**Service Attendant**	J. HARRIS MOBIL SERVICE STATION, Emeryville
1986–87	**Carpenter Helper**	HOYER TERMITE CO., El Cerrito

See Tricia's cover letter on page 264.

TRICIA BAKER

9490 Wheelwright Road • Clio, MI 48409 • (810) 876-5432

Objective: A position in Sales/Management, dealing with consumer services or high quality consumer products.

SUMMARY

- Experienced – Ten year track record of success in sales.
- Responsible – Continually searching for more responsibility.
- Efficient – Expert in time management.
- Influential – Able to encourage people to produce.
- Versatile – Able to handle a variety of tasks.

WORK HISTORY

APEY DECAL FACTORY, ALDERSON, MI	***Graphic Artist***	1992–present
NORMAN WILLIAMS COSMETICS, SAGINAW, MI	***Manager***	1991–92
NORMAN WILLIAMS COSMETICS, FLINT, MI	***Manager***	1982–90

PROFESSIONAL EXPERIENCE

Management

- Supervised staff of fifteen:
  - Interviewed, hired and terminated staff;
  - Managed, motivated and trained staff; resolved communication problems;
  - Scheduled work shifts, processed payroll, wrote monthly evaluations.
- Promoted to management position after one year in the business. Held position for the next eight years with continual raises and bonuses for an outstanding job.
- Helped employees earn higher wages by teaching them to manage their time.
- Missed less than 10 days of work over the past thirteen years.
- Increased overall sales by 20% within a nine month period upon transfer to a new location. This generated higher profits as well as benefits for employees.

Sales

- Consistently exceeded monthly and annual sales goals, receiving highest sales award every month for over 5 years.
- Developed an expertise in customer service and "closing the sale."
- Experience in opening/closing cash register, counter sales, telephone follow-up.

Purchasing

- Increased company profits by finding products at lower prices, successfully anticipating customer needs and new trends.
- Handled monthly inventory, purchasing, and bookkeeping.

Computer Skills

– WordPerfect	– Aldus Freehand	– Lotus 1-2-3
– O-Foto	– Macintosh Desktop Publishing	

– Resume written by the job hunter –

Judy's work history is complex and needed a lot of explaining. This was our best compromise.

JUDY ROGERS
392 St. George Avenue, #11
Alameda, CA 94501
(510) 335-2445

Objective: Sales/Marketing Position.

QUALIFIED BY

- Over 15 years' professional experience with the public.
- Personable and persuasive in communicating creatively with thousands of customers from all cultures and economic levels.
- Proven skill in persevering to solve customers' problems.
- Self-motivated and confident in making independent decisions.
- Very well organized and able to meet deadlines.

RELEVANT EXPERIENCE

Sales & Marketing

- Made direct presentations to Bay Area retail store owners and buyers, marketing Christmas ornaments and gift items imported from the Philippines.
- Co-hosted sales seminars for potential real estate partnership investors:
  – Oriented customers by answering questions regarding project details;
  – Followed up by phone to verify their commitment to invest in the partnership.
- Canvassed by cold calling for contributions to a nonprofit organization.
- Consistently surpassed sales quotas in Macy's clothing and houseware departments.

Customer Service

- Resolved wide range of customer problems, applying diplomacy and assertiveness to: delivery delays; fee and budget problems; property management decisions; airline emergencies and in-flight problems; and culture/communication barriers.
- Organized the logistics of speaking engagements and investment seminars:
  ...location ...catering ...seating ...literature ...speakers ...travel ...RSVP calls.

EMPLOYMENT HISTORY

1995–present	**Bookkeeper**	OLSON LIGHTING CONSULTANTS, SF
1993–95	**Office Manager**	GROTHE REAL ESTATE, San Francisco
1990–92	**Philippine Import Sales***	SELF-EMPLOYED, selling to Bay Area stores
1988–90	**Neo-Life Vitamin Sales***	SELF-EMPLOYED, selling to flight attendants
	*(part-time, concurrent with airline employment)	
1981–92	**International Flight Attendant**	TRANSAMERICA AIRLINES, Oakland
1980–81	**Editorial Coor./Sales Sec'y.**	PSYCHOLOGY TODAY Textbook Div., San Diego
1975–79	**Retail Sales**	MACY'S, ROOS ATKINS, part-time during college

EDUCATION

B.A., Speech/Theatre Arts – UC, Santa Barbara

LINDA MOWRY

348 Somerset Road
Hayward, CA 94541
(510) 666-7995

Objective: Position as Sales Coordinator, Sales Rep or Account Executive.

SUMMARY

- 10 years' successful experience in direct sales.
- Enthusiastic and motivated; sincerely enjoy developing and maintaining good client relations.
- Professional in appearance and presentation.
- Effective working alone and as a cooperative team member.

EMPLOYMENT HISTORY

1995–present **Sales Coordinator** JANA IMPORTS
Imported giftware, Oakland

- Promoted giftware products at trade shows, greeting new and old customers.
- Coordinated product information and distribution for 75 field reps and major accounts.
- Developed mock-up and organized details for effective photography for 20-page catalog.
- Maintained current inventory status reports, summarizing computerized data.
- Coordinated shipping documentation for imported products, and maintained files on foreign manufacturers and custom brokers.

1992–94 **Distribution Coordinator** DEJA VU PUBLISHING CO.
Magazine publisher, San Rafael

- Promoted and developed new distribution outlets for this special interest magazine:
  - Made cold-call and follow-up visits to retail outlets throughout Northern California;
  - Organized and maintained detailed routebooks and all related financial records;
  - Succeeded in increasing readership by 40% over a two-year period.
- Increased advertising revenue by researching publications and by bulk-mail promotion.
- Coordinated production of advertising to appear in major trade publications.

1986–91 **Co-owner/Manager** BILL'S DAIRY PRODUCTS
Retail/wholesale milk, Livermore

- Sold dairy delivery services to both retail and wholesale customers, as co-owner of this small business involving a staff of five.
- Handled all aspects of order taking and order processing.

EDUCATION

Health studies, BAY CITY COLLEGE, San Francisco

SALES

MARK S. FLEETWOOD

1299 Bruenner Avenue
Castro Valley, CA 94546
(510) 335-9009

Objective: Position as Electronic Sales Representative.

SUMMARY

- Number One ranked salesman for 4 straight years.
- Strong product knowledge from 5 years' experience in the field.
- Enthusiastic about both the product and the role of sales rep.
- Able to handle large territories effectively.
- Experience serving wide range of electronics industries.

PROFESSIONAL SALES EXPERIENCE

Account Relations/Customer Service

- Established and maintained good rapport with over 200 accounts in the electronics service industry:
  - Followed through promptly to resolve customer complaints;
  - Found hard-to-find parts for customers, by whatever means necessary;
  - Located detailed product info for customers to facilitate accurate parts ordering.
- Currently servicing retail and wholesale accounts, visiting each account weekly.

Direct Sales & Product Demonstration

- Set sales record, surpassing all salesmen for any given month in company history.
- Held down company's largest territory; exceeded quotas and greatly increased sales.
- Increased average monthly sales to Pacific Stereo from $2,500 to $21,000.

Product Analysis & Forecasting

- Reviewed potential new products applicable to electronics industry:
  - Tested products on various applications;
  - Introduced products to customers for their evaluation.
- Projected likely success rate of new items, computing results of customer surveys conducted by phone and mail.

Marketing/Promotion

- Effectively demonstrated to customers the benefits of quantity purchases and incentive programs.
- Organized customized accessory racks for display in retail outlets.
- Researched industry trade journals to identify potentially popular and profitable items, and locate products requested by accounts.

EMPLOYMENT HISTORY

1992–present	**No. Calif. Field Sales Rep.**	PARNASSUS ELECTRONICS, Oakland, CA
1990–91	**Full-time student**	DIABLO VALLEY COLLEGE
1989	**Warehouseman**	BILLINGS PRECISION box mfg., San Leandro, CA
1985–88 summers	**Warehouseman**	BILLINGS PRECISION box mfg., San Leandro, CA

EDUCATION

Business Administration & Sales, Diablo Valley College, Pleasant Hill, CA 1990-91

MELINDA SAILOR

222 Billings Drive ● San Francisco, CA 94132 ● (415) 122-6795

Objective: Position in fine jewelry sales and customer service
with a quality jewelry store or contemporary design gallery.

PROFILE

- Successful experience in sales and customer service.
- Knowledge of fine gems and jewelry.
- Widely varied, well rounded background in art and design.
- Excellent communication skills; fluent in Spanish.
- Sincere commitment to professional growth in the field.

EMPLOYMENT HISTORY

1996–present **Preview Staff** RICHARDS Auctioneers/Oakland
- Set up jewelry displays for Richards' Jewelry Department previews.
- Learned to identify a wide variety of gems, pearls and precious metals, from direct experience handling and cataloguing gemstones.
- Assisted gemologist in documenting appraisals of estate and period jewelry.

1995 **Exhibit Technician** TRITON MUSEUM OF ART, Santa Clara, CA
- Prepared displays and fine arts exhibits at local museums.
- Entered data on computer for catalogue descriptions written with specific gemology terms of the industry.

1994–95 **Sales Clerk** EMPORIUM CAPWELL, San Francisco
- Successfully developed a combination of retail selling skills:
  - Greeted customers and determined their specific needs;
  - Handled problem customers with patience and sensitivity;
  - Utilized "suggestion selling" techniques leading to effective closing;
  - Followed up, encouraging customers to return, generating repeat business.
- Demonstrated jewelry and advised about style;

1991–94 **Sales Associate** MACY'S OF CALIFORNIA – San Mateo, CA
- Maintained detailed paperwork for retail sales inventory control.

1989–91 **Teacher**, art & adult ed. Los Angeles, CA and Miami, FL
- Developed a keen appreciation for basic design principles, as a student and teacher in the art field. (Professional classes listed below.)

1986–88 **Teacher**, Arts & Crafts PEACE CORPS – El Salvador, Central America

EDUCATION & TRAINING

B.A., Teaching Credential – Fine Arts; Immaculate Heart College, Los Angeles, 1985

Certificate for completion of professional jewelry workshops:

• Design & Rendering • Repair & Fabrication • Basic Goldsmithing

NOREEN MacLAUGHLIN
1201 Genner Street, Apt. 12
San Francisco, CA 94111
Home: (415) 577-0121 Business: (415) 439-6897

OBJECTIVE

Professional sales position.

SUMMARY

- Confident, poised, well presented and competent sales professional.
- 15–year successful record in sales, sales training, and sales management.
- Outstanding communication / listening skills, achieving client confidence.
- Record of high profitability in sales and marketing promotions.
- Strong decision-maker; goal- and profit-oriented.

PROFESSIONAL EXPERIENCE

1982–present	**Assistant Administrator**	HILLARY APARTMENTS – San Francisco
"	**Director of Resident Relations**	" "
"	**Sales Supervisor**	" "

- Created sales and marketing programs that increased residential rent revenue by 20% per unit, and increased shopping center profits by 33%.
- As Director of Resident Relations, successfully increased resident retention rate through greatly improving customer relations and applying acute listening skills.
- Developed strong pattern of repeat sales and client loyalty; provided accurate and honest product information, and help in assessing client needs.
- Consistently surpassed companies' sales records; promoted to sales management.
- Interviewed and hired top sales people, achieving low turnover.
- Trained over 25 assistants and sales personnel:
  – Created a safe, comfortable learning environment with constructive feedback, resulting in professionally trained staff;
  – Taught mastery of effective basic selling techniques:
  ...selling benefits ...handling objections ...cold calls ...demonstrations ...closing.

1981 **Sales Director** GILBERT GALLERY – Ghirardelli Sq., SF

- Made presentations to merchants on the use of print, radio, coupons and raffles.
- Conducted cold calls to business neighbors of gallery, significantly increasing business and gallery profits.

EDUCATION

B.A. Humanities – San Francisco State University
Humanities – UC Berkeley
Nursing – Niagara University

Professional Associations:
National Association of Professional Saleswomen; Northpoint Merchants Association

– References available upon request –

SALES

PAMELA SWISS

1930 Atlantic Avenue
San Francisco, CA 94115
(415) 688-0131

OBJECTIVE

Position as outside sales representative for a manufacturer, specializing in fashion, cosmetics, and / or accessories.

SUMMARY

- Strongly self-motivated, enthusiastic and profit oriented.
- Outstanding communication and presentation skills.
- Readily project a professional, fashionable image.
- Extremely sociable, able to put clients at ease.
- A decision maker; well organized, resourceful, and work well independently.

RELEVANT EXPERIENCE

Sales, PR, Marketing

- Sold advertising to major SF corporations for a special newspaper supplement in support of the SF symphony:
  – Coordinated with other sales team members; – Met deadlines;
  – Worked independently; – Sold to influential business leaders.
- Represented United Airlines at public relations events.
- Demonstrated manufacturers' products at trade shows and conventions, as independent interior designer.

Client Services

- Supervised direct services to thousands of airline clients, assuring that individualized needs were met, remaining calm and effective under stressful conditions.

Fashion & Design

- Conceptualized and implemented unique and artistic contemporary designs for residential environments:
  – Incorporated the client's personal preferences and tastes;
  – Researched products available, consistent with client's budget and priorities;
  – Collaborated with architects, construction subcontractors, and upholsterers;
  – Prepared and submitted detailed monthly reports to clients on budget and expenses.
- Studied Interior Design at Cañada College.
- Completed courses in Fashion Coordination, Self-image, Basic Elements of Design.
- Advised businesswomen on fashion coordination, image development and personal style.

WORK HISTORY

1986–present	**Flight Attendant**	UNITED AIRLINES, S.F.
1985	**Marketing Asst.**	S.F. SYMPHONY (Volunteer)
1983–85	**Owner/Designer**	PAMELA MARSH INTERIOR DESIGN, S.F.

EDUCATION

Liberal Arts & Design, CALIF. STATE UNIVERSITY, San Luis Obispo
Interior Design, CAÑADA COLLEGE, Redwood City

SHERRIE E. VALENCIA

378 Cornell Street - Apt. 409 ◆ San Francisco, CA 94109 ◆ (415) 178-2323

Objective: Position in sales/marketing.

PROFILE

- Over 10 years' experience in sales and account management.
- Honest; straightforward; respected and trusted by clients.
- Effective working in self-managed projects and as member of a team.
- Outstanding communication, analytical and presentation skills.
- Sharp, innovative, quick learner; proven ability to adapt quickly to a challenge.

SALES / MARKETING EXPERIENCE

1990 – present **District Sales Manager** L.B.HEINE TRUCKING CO., Hayward

Marketing, Sales Presentation

- Planned successful strategies to target and develop new accounts.
- Consistently expanded customer base by at least 50%, and increased revenues from current clients by 25%.
- Made oral presentations to upper management of major corporations, such as Bechtel, Union Carbide, Chevron, Lockheed.

Planning/Organizing

- Assessed and evaluated market conditions to identify sources for potential new client base.
- Developed and revised daily, weekly and monthly plans of sales strategies.
- Organized Northern California territory to maximize efficiency of calling pattern.

Report Writing

- Wrote timely reports and forecasts to management on past and projected sales volume; wrote evaluations, problem analyses, and daily plans.

1988 – 89 **Accounts Executive** C.F. AIR FREIGHT CO., South San Francisco

- Restored and maintained good working relations with clients:
  - – Maintained daily telephone contact with current accounts;
  - – Made field visits and discussed customers' problems;
  - – Researched problem areas and provided detailed information;
  - – Followed through quickly and thoroughly with satisfactory resolutions.

1986 – 87 **Accounts Executive** EAST TEXAS MOTOR FREIGHT, So. SF

1985 – 86 **Sales Representative** AMERICAN INDUSTRIES INC., SF

EDUCATION

B.S., Education – Ohio State University
Sales/Marketing courses, UC Berkeley; City College of SF; SF State University

Her earlier work and business degree from Switzerland are mentioned, without going into detail.

VRENY ZURICH

314 Glenellen Avenue
San Francisco, CA 94118
(415) 555-2331

OBJECTIVE

Position as small-store manager, department manager, or salesperson, preferably dealing with quality shoes, handbags and accessories.

SUMMARY

- 18 years' successful experience in shoe sales.
- Extremely reliable, hard working, and honest.
- Establish excellent relations with customers, building loyal repeat business.
- Work well in a team with people of all ages.
- Well organized and thorough in completing projects.

SALES EXPERIENCE

DIRECT RETAIL SALES

- Successfully sold high-fashion, high-quality men's and women's shoes at world-famous shoe stores: Joseph's, Charles Jourdan's, and Denver's, in San Francisco.
- Achieved position of top salesperson at Charles Jourdan Boutique.

CUSTOMER SERVICES/CUSTOMER RELATIONS

- Developed loyal customer base and increased sales volume through personal attention to customers:
  - Maintained detailed record of individual customers' buying habits and preferences;
  - Contacted customers to notify them of special sales and new merchandise shipments;
  - Sent thank-you notes for patronage, and cards on special occasions.

MANAGING/ORGANIZING

- Consistently maintained clean, attractive shopping area and well organized stockroom, assuring that merchandise was accurately replaced in stock room.
- Managed supermarket near Zurich, Switzerland, supervising and training 15 employees, filling in at all positions, and handling store opening, bookkeeping, buying and promotion.

WORK HISTORY

1995–present	**Salesperson** – shoes	DENVER'S FINE SHOES – San Francisco
1985–94	**Salesperson** – shoes	CHARLES JOURDAN – SF
1978–85	**Salesperson** – shoes	JOSEPH'S SHOE BOUTIQUE – SF

Additional previous store management in Switzerland.

EDUCATION & TRAINING

Degree in Trade and Business, Switzerland

Eleanor inserts a few skill-headings in her chronological resume to emphasize skills related to the job objective.

ELEANOR T. KENNEDY

809 Laurel Drive • San Francisco, CA 94123 • (415) 202-6164

Objective: Position as Sales Representative.

PROFILE

- Experience in sales, handling entire western region of U.S.
- Enjoy a challenge; work well under pressure.
- Self-motivated, goal-oriented and well organized.
- Readily inspire the confidence and trust of customers.

PROFESSIONAL EXPERIENCE

1995–present **Western Regional Sales Representative**
COMPUTER PRODUCTS FOR THE BLIND, Redwood City, CA

Sales Management

- Managed and operated western regional office:
  - Resolved customer problems, serving as liaison with the home office;
  - Sold directly to major corporations such as Hewlett-Packard, McDonnell-Douglas, Hughes Aircraft, Los Angles Times;
  - Developed extensive database of potential clients through contacting agencies, educators, vocational rehabilitation administrators, corporate managers;
  - Followed up on referrals and leads, and established new customer contacts, to arrange product demonstrations.

Direct Sales

- Demonstrated and sold a wide range of products for the blind and visually impaired:
  ...speech output devices ...braille printers ...electronic mobility aids.

1993–94 **Outside Sales Consultant**
TRAVEL DESIGN, INC. (part time); Mountain View, CA

- Made cold calls and field visits to new customers, significantly increasing accounts.
- Sold travel packages to a broad range of clients acquired through individual contact, coordinating vacation itineraries, reservations and ticketing.
- Planned and implemented individualized travel skills programs for low-vision and blind veterans, including their family members.

1991–93 **Blind Rehabilitation Specialist**
VETERANS ADMINISTRATION MEDICAL CENTER; Palo Alto, CA

Presentation & Training

- Delivered group presentations at national and regional conferences, addressing professionals in the field of blindness.
- Designed and presented workshops throughout the Western U.S., training special educators, vocational rehabilitation counselors, administrators and individuals, on the use of computer products for the blind.

1985–90 Full Time Student
" " **Progress Specialist** (summer) EASTER SEALS FOUNDATION; Inkster, MI
" " **Park Technical** (seasonal) ROCKY MTN. NATIONAL PARK, CO

EDUCATION

M.A., Blind Rehabilitation – WESTERN MICHIGAN UNIVERSITY, Kalamazoo, MI
B.S., Therapeutic Recreation – EASTERN MICHIGAN UNIVERSITY, Ypsilanti, MI

SALES

A Potpourri of Resumes

• Environmental

Barbara McClosky — Info specialist/environmental 200
Brian Rebar — Environmental technician 201

• Product Development

Emily George — Clothing designer/illustrator 202
Joyce Stroebech — Production assistant, clothing 204
Rebecca Vaness — Production assistant, clothing 205
Wendy Gillroy — Product developer/client services rep 206

• Film, Radio, TV

Katherine Brunswick — Film production assistant 207
Lynn Sheffield — Radio or TV programming/announcing 208
Martha Azcona — TV programming production 209
Rebecca Newburg — Film production manager/assistant 210
Roger Lancaster — Entry level, TV news grip 211

• Property Management

Adrienne Mendoza — Property management, entry level 212
Mark Killorin — Property management trainee 213
Richard Flores — Property management, Wells Fargo 214

• Graphic Design

Amy Buchannon — Graphic designer assistant 215
Barbara Monet — Graphic production 216
Gregory Byron — Entry level, multi-image production 217
Lynne Charney — Design/production, publications 218
Vickie Wan — Design, multi-media 219

• Legal

Ann Voorhees — Staff attorney 220
Michael Oliver — Law office clerk, research assistant 222
Randolph Strough — Paralegal 223
Andrew Thompson — Legal assistant 224

• Editing/Writing

Richard Griffon — Freelance editing, proofreading 225
George Amundsen — Editorial assistant, Chevron 226
Rebecca Bridges — Editorial assistant, publishing 227
Mary Moriarity — Freelance book editing 228
Carolyn Clarke — Editorial assistant, book publishing 230

• Real Estate Appraisal

Linda Durkee — Real estate appraiser, entry 231
Margaret Lester — Real estate appraiser trainee 232
Dennis Brinkley — Real estate appraisal trainee 233

• Religion

Susan Holmes — Pastoral minister 234
Sister Mary Jones — Pastoral associate 236

• One-of-a-Kind

Jo Anne Burgess — Union rep/business agent 238
Bradley French — Writer/photographer/editorial assistant 239
Loren Greene — Summarizer, Barron's legal services 240
Michael Wong — Community or government relations rep 241
Richard Jennings — Fire fighter, entry level 242
Robert Lawton — Service writer/auto manufacturer 244
Stephen Honda — Real estate analyst/researcher 245
Donald Raulings — Private investigator 246
Teresa Fernandez — Wardrobe assistant, movie/TV 248
Kenneth Richards — Warehouseman 250
Carlene Doonan — Apprentice baker 251
Dawn Ellsworth — Gallery assistant/sales 252
Hellmut Dietrich — Freight handling, import/export 253
Dolores Walker — Commercial leasing, agent 254
Denise Francis — Activity director, cruise staff 255

BARBARA McCLOSKY

1721 Washington Street, Apt. 2
Ann Arbor, Michigan
(313) 521-7080

OBJECTIVE

Information specialist, resource manager, or comparable position with an environmental consulting firm.

SUMMARY

- Ten years' professional experience in information management.
- Special talent for quickly and accurately locating needed information.
- Successfully designed and created an Environmental Resource Library.
- Degrees in Librarianship and Conservation Resource Studies.

PROFESSIONAL EXPERIENCE

INFORMATION NEEDS ANALYSIS/ADVISING

- Matched up people and information, providing faculty, students and professionals from a broad range of disciplines, with specific resources:
  - Books on resource management, environmental policy, development, assessment and methodology;
  - Audiovisual materials: audio tapes, video tapes, and slides on environmental topics;
  - Bibliographies, periodicals;
  - Reports of governmental agencies, educational institutions, and nonprofit organizations.
- Developed highly valued personalized networking and information service, known for reliability, resourcefulness and helpfulness; provided job referrals, internships and volunteer placements, and leads on experts in the field.

RESEARCH & WRITING

- Researched and authored *Library Guide to Information on the Environment.*
- Co-authored *Students' Guide to Audiovisual Materials on Campus.*
- Wrote and edited campus and alumni newsletters on information resources and environmental issues.

DATA MANAGEMENT

- Conducted thorough assessment of information systems of other libraries, and incorporated the most appropriate features of each.
- Designed specialized catalogs and indexes for quick access to a variety of research materials.
- Continually revised and upgraded library systems to meet changing information needs.

WORK HISTORY

1995–present	**Assistant Director**	Natural Resources Intern Program, Ann Arbor, MI
1982–94	**Director**	Environmental Resource Center, Department of Environmental Studies, University of Michigan

EDUCATION

B.S., Environmental Studies – UNIVERSITY OF MICHIGAN, Ann Arbor, MI
M.A., Librarianship – CA STATE UNIVERSITY, San Jose, CA

POTPOURRI

Brian is just completing his AA degree and has no paid work experience in the field, so he describes his recent training in detail. Brian sent out 10 copies of his resume and got 4 telephone interviews in the first week.

BRIAN M. REBAR

1250 Skipper Road, Apt. 38, Tampa, FL 33613 • (813) 977-7558

OBJECTIVE: Seeking a career in the Environmental Field

EDUCATION:

Specialized Associate Degree in Environmental Technology — Anticipated May 1996
ENVIRONMENTAL TECHNOLOGY CENTER, Tampa, FL — Current GPA: 3.7 / 4.0

Areas of Study:

- Surface Water Sampling
- Ground Water Sampling
- Technical Support Skills
- Air Quality Sampling
- Hazardous Waste Operations
- Wetland Management
- Underground Storage Tank Management
- Soil Sampling

Field Work and Studies:

- Assisted a FDEP biologist with a mollusc study conducted on the Myakka River.
- Field trained in Surface and Groundwater sampling and data collection.
- Familiar with decontamination protocol of equipment (FDEP guidelines).
- Determined Groundwater flow using a transit.
- Experienced in well development, purging and installation.
- Field trained in soil logging and identification using the Munsell Chart.
- Field trained in Wetlands mitigation and delineation using state method (62-340 FAC).
- Performed Phase I Risk Assessment and Phase II Constraints Analysis.
- Interpreted serial photography and legal descriptions.
- Performed records reviews (FDEP, EPC, EPA).

Equipment Used:

- Kemmerer Stainless Steel Sampler
- Van Dorn Teflon Sampler
- Stainless Steel and Teflon Boilers
- American Sigma Water Sampler (3700)
- Hester-Dendy Substrate Sampler
- Global Positioning System (GPS)
- Level A, B, and C Safety Equipment
- Bladder Pump
- Hydrolab
- Ponar Dredge
- Isco Portable Sampler
- Isco Flow Meter
- Transit
- Handheld Flowmeter
- SCBA
- Peristaltic Pump

Computer Skills:

- IBM Compatible PCs
- Quattro Pro
- Windows & DOS
- Some AutoCAD
- WordPerfect

Certifications:

- 40 Hour HAZWOPER Certificate (OSHA CFR 1910.120(e) 3)
- American Red Cross First Aid and Adult CPR

ST. PETERSBURG JR. COLLEGE, Clearwater, FL — 1991–94

- 39 Credit Hours Completed

WORK HISTORY:

Crew Leader, TACO BELL, Clearwater, FL — 1989–95

- Oversaw crews of 3 to 8 people.

Stock Room Inventory Control, ORGANIZER DEPOT, Clearwater, FL — 1988–89

- Maintained the stock room.

– Resume written by Brian M. Rebar & Nancy Rosenberg –

EMILY GEORGE
2440 Parker Street
Berkeley, CA 94704
(510) 544-2378

Objective: Position as designer or illustrator, working with clothing, jewelry, interiors, graphics, or displays.

SUMMARY

- Creative artist with 12 years' experience in clothing design.
- Successfully designed and marketed Emily George Artwear, my own line of high-fashion women's clothing.
- Sharp, well organized, hard working; able to meet deadlines.
- Dedicated to highest quality work.
- Able to manage and implement a project from initial design concept through final production.

PROFESSIONAL EXPERIENCE

Product Design

- Designed appliquéd and hand-painted women's jackets for quality clothing market, including: – Neiman Marcus/Beverly Hills – Liberty House of Hawaii – White Duck – Marriott Hotel Gift Shops – La Costa Resort – Maison Mendessolle
- Exhibited and sold original clothing and accessory designs at:
  – American Craft Council Crafts Fair, San Francisco, 1995;
  – Palo Alto Celebrates the Arts, 1995;
  – Pacific Fashion Institute Benefit Fashion Show, 1993–94 (by invitation);
  – Oakland Museum Natural Crafts Fair, 1992–93;
  – Oakland Museum History Guild Needlework Exhibition, 1995.

Graphic Design

- Designed advertising mailers for my clothing design business: directed photo sessions; designed layout; wrote copy; selected typestyle and paper; coordinated with printer.

Display, Merchandising, Sales

- Designed and built convertible and portable display booth for craft fairs:
  – Prepared scale drawings; purchased fabric, wood and hardware materials;
  – Supervised construction and painting; arranged merchandise display.
- Created window displays and in-store displays for The Soft Touch Boutique.
- Set up special displays of new and sale merchandise in working studio setting.
- Developed and maintained customer mailing list.
- Negotiated with manufacturer's reps to handle my designs as independents, and at Los Angeles Fashion Mart.

– Continued –

PROFESSIONAL EXPERIENCE, continued

Materials: Planning & Purchasing

- Researched sources and bought fabrics and notions at trade shows, warehouses, specialty wholesalers, and through manufacturers' representatives.
- Coordinated the selection of fabrics and findings, and the customizing of materials, for each season's line.
- Budgeted materials purchase based on projected season sales and calculations of quantities needed.
- Established business credit with manufacturers.

EMPLOYMENT HISTORY

1986–present	**Designer/Owner**	EMILY GEORGE ARTWEAR, Berkeley
1984–85	**Partner/Designer**	THE SOFT TOUCH BOUTIQUE, Lafayette
1981–83	**Clothing Designer**	Self-employed, custom/contract design and sewing
1979–81	**Library Assistant**	UC BERKELEY, Bancroft Library
1977–79	**Musician/Violist**	SANTA BARBARA SYMPHONY ORCHESTRA
1977–78	**Retail Sales**	ROBINSON'S DEPT. STORE, Santa Barbara

EDUCATION & CREDENTIALS

B.A., Music; performance emphasis: Viola; UC SANTA BARBARA, 1979
Private Pilot's License – Oakland, 1989

ADDITIONAL STUDIES

CCAC Extension: Life Drawing, 1992; Botanical Illustration; 1994, Painting & Color, 1994
Pacific Basin School for the Textile Arts: Beginning & Intermediate Silk Screen, 1995–96
Pacific Fashion Institute, 1987: Pattern Drafting; History of Costume; Fashion Illustration
Fiberworks: Color Theory, 1988; Soft Sculpture, 1989
UC Extension: Small Business Management, 1986

See Joyce's cover letter on page 267.

JOYCE STROEBECH
578 Willow Drive
Walnut Creek, CA 94598
(510) 902-1228

Objective: Position as Product Assistant for KARAN Accessories.

PROFILE

- Enthusiastic team member whose participation brings out the best in others.
- Production experience in manufacturing textile products, including purchasing materials, pricing products and inventory planning.
- Resourceful problem solver who is good at details.
- Proven success at developing new products with broad market appeal.
- Highly committed to the KARAN philosophy and eager to support it.

RELEVANT EXPERIENCE

Production and Design

- Developed and produced fabric accessories, children's clothing and soft sculpture toys for retail market:
  - – Purchased raw materials;
  - – Determined retail and wholesale prices of products;
  - – Processed purchase orders and invoices;
  - – Shipped and delivered merchandise, meeting deadlines;
  - – Maintained inventory and kept bookkeeping records.
- Ordered and distributed wholesale fabric and notions for 150-member textile artists' guild; set up and kept bookkeeping and tax records; processed invoices.
- Controlled inventory, collected sales records, received merchandise, and processed invoices for retail store, as manager of Fyneline Gifts, Inc.

Marketing and Sales

- Established national market for my original gift line, expanding wholesale accounts from 4 to 22 stores.
- Increased sales in several departments of Fabric Heaven store by combining merchandise to create fashion jewelry display.
- Sold clothing, gifts, cosmetics and fabrics at retail level.
- Developed good client relations on both wholesale and retail levels.

Management

- Managed Fyneline Gifts and supervised eight employees.
- Administered wholesale buying co-op for 150-member textile artists' guild.
- Increased sales and variety of merchandise at Fabric Heaven.

EMPLOYMENT HISTORY

1986–present	**Fabric & Craft Sales**	FABRIC HEAVEN, Pleasant Hill
1979–present	**Designer/Owner/Wholesaler**	FIBER DESIGNS, Walnut Creek
1976–78	**Assistant Cosmetician**	RUNYON'S DRUG, Walnut Creek
1974	**Assistant to Decorator**	MOBILIA FURNITURE, Houston, TX
1969–71	**Gift Shop Manager**	FYNELINE GIFTS INC., Warren, OH

EDUCATION

B.A., Home Economics: Interior Design & Textile Art – UNIVERSITY OF HOUSTON

REBECCA VANESS

2707 Benvenue Avenue
Oakland, CA 94606
(510) 313-0854

Objective: Position as Production Assistant with a clothing manufacturer, wholesaler or design house.

SUMMARY

- ✦ Experience with all aspects of garment production.
- ✦ Hands-on knowledge of sewing and pattern development.
- ✦ Effective and experienced sample room supervisor; special talent for bridging language barriers.
- ✦ Proven ability to assume increasing responsibility.
- ✦ Committed to maintaining exacting standards.

RELEVANT EXPERIENCE

Production/Engineering

- Coordinated sample room activity:
  - Maintained wide variety of sample yardage to meet current and ongoing needs of merchandisers and engineering department;
  - Monitored and replenished supplies;
  - Maintained records of all clothing lines in production for referral of merchandisers.

Supervision

- Balanced work flow of pattern makers, sample makers, cutter, and merchandisers, according to established deadlines and production considerations.
- Promoted cooperative and productive working environment:
  - Assessed needs and abilities of each person;
  - Maintained an overview of each project;
  - Supported staff by assuming appropriate responsibility.
- Negotiated and established realistic deadlines.

Quality Control

- Inspected garments, checking for:
  ...correct seam construction, finish stitching and trims ...appropriate packaging
  ...fabric quality.
- Measured first-production and counter samples, comparing with specifications.

WORK HISTORY

1995–present	**Design Room Assistant**	BAY CITY UNIQUE CREATIONS (private label design firm)
1989–94	**Freelance Seamstress**	Self-employed
1987–89	**Cutter/Seamstress**	DAY'S CUSTOM UPHOLSTERY, Spokane, WA
1985–87	**Seamstress/Designer**	Freelance, concurrent w/travel in Europe
1984–85	**Boutique Seamstress**	NAPOLEON & JOSEPHINE'S boutique, Berkeley

She explains what "Bay City" is about.

Language skills: familiar with Spanish, French, Greek, German

WENDY V. GILLROY

900 Valleyview • Sausalito, CA 94965
(415) 901-2662

OBJECTIVE

Project coordinator, product developer, or client services representative, with a local manufacturer.

PROFILE

- Demonstrated success in planning and completing large, complex projects.
- Technical expertise in the applications of manufacturing materials.
- Exceptionally creative in design, utilization and problem solving.
- Outstanding skill in researching all aspects of material and design data.
- Effective in public presentation of products and services.

EMPLOYMENT HISTORY

1994–present **Project Manager** CITY SAVINGS BANK – San Francisco

- Served as owner's representative on multi-million dollar construction projects for City Savings:
  - **Initiated specifications** and **developed contracts** for each project;
  - **Developed project budgets and schedules**; coordinated tenant leasing;
  - Supervised and **coordinated all construction** and administration matters.

1986–93 **Owner/Manager** GROUNDS FOR PLAY, INC. – Atlanta, GA

- **Developed promotional and advertising materials** for my own design and construction business.
- Consulted with clients to **assess their design wants and needs**; then **produced custom designed furniture** and individualized interiors.
- **Developed unique play environment** for handicapped children, based on their physical skills and limitations, and incorporating **innovative uses of conventional materials**.
- Researched and studied technical properties and applications of many construction materials:
  ...metals ...wood products & preservatives ...plastics & vacuum forming.
- **Designed prototype playground** for installation in 14 public parks in Atlanta, GA. Developed manufacturing schedules, coordinated subcontractors and **completed all 14 installations in four months**.
- **Delivered informational presentations** on community resources, using slide shows and talks.

1985–86 **Art Critic** ATLANTA GAZETTE – Atlanta

EDUCATION

B.A., Art History – Goucher College, Baltimore, MD
B.F.A., Sculpture – Atlanta College of Art, Atlanta, GA

Katherine organizes her accomplishments by skill/media, yet makes it very clear where things happened.

KATHERINE BRUNSWICK

4980 - 16th Street, Apt. 6 • San Francisco, CA 94114 • (415) 151-5887

Objective: Position as Production Assistant with film company.

QUALIFICATIONS

- Production Assistant for television series.
- Production Coordinator for independent film company.
- Assistant to Producer, Mill Valley Film Festival.
- Experience in radio production at university radio station.
- Acted in, directed and co-authored short 8mm film.
- Degree in Art History, with emphasis in art and graphics.
- Lifelong exposure to film and television industries.

RELATED EXPERIENCE

FILM • TELEVISION

Silver Productions, ABC-TV:

- Assisted in office supervision for producer and production coordinator.
- Coordinated transportation for cast and crew members.
- Oversaw communication between Oakland and LA offices.
- Assisted Casting Director and Location Manager.

Farquart Film Productions, as Production Coordinator:

- Scouted locations, acquired filming permits, obtained props.
- Organized catering, chauffeuring, flight bookings, and accommodations.
- Supervised crew, talent, musicians, extras, clientele.
- Wrote and distributed press releases. • Photographed promotional stills.
- Assisted Art Director; designing/building sets, lighting pick-up shots.

FILM-RELATED

At Mill Valley Film Festival, as Production Assistant:

- Developed and coordinated special children's film series.
- Organized and assisted with operational tasks of film institute office.
- Planned and implemented operational model for Special Events Department.

COMMUNICATION • MEDIA

At KALX Radio, as news writer, anchor and reporter:

- Engineered and anchored a weekly newscast.
- Investigated and reported on student rallies and campus activities.
- Taped live, on-air radio reports of the 1984 Democratic National Convention.
- Trained and supervised incoming staff members.

EMPLOYMENT HISTORY

1996–present	*Production Asst.*	SILVER PRODUCTIONS, ABC-TV – Oakland, CA
Fall 1995	*Production Coord.*	FARQUART FILM PRODUCTIONS – Berkeley, CA
Fall 1995	*Production Asst.*	MILL VALLEY FILM FESTIVAL – Mill Valley, CA
1994–95	*Producer*	KALX Radio – UC Berkeley
1991–95	*Full-time student*	UC Berkeley and University of Oregon

EDUCATION & TRAINING

B.A., Art History - University of California, Berkeley, 1995

Universita di Firenze - Florence, Italy, 1992

Lynn also has both M.A. and Ph.D. degrees, but we left them off since it would actually detract from her current goal and appear inconsistent with her work history.

LYNN M. SHEFFIELD

6023 Dover Street, #B
Oakland, CA 94609
(510) 699-8000

Objective: Position with a radio or TV station in programming and/or announcing.

PROFILE

- Wide experience in the performing arts; successfully wrote and recorded songs.
- Resonant speaking voice; accomplished singer.
- Excellent writing skills in both music and lyrics.
- Lifelong commitment and involvement with music.
- Sociable, personable, communicate easily with a wide range of personalities.

PROFESSIONAL EXPERIENCE

Creativity & Performing Skills

- Successfully wrote and performed four songs on a major record label, as member of ISIS, a pioneering and nationally recognized women's band. ("Breaking Through," United Artists)
- Recorded extensively with other professional musicians:
  - Sang (and performed in concert) with David Amram, classical and jazz composer, recording "No More Walls" on RCA Records;
  - Sang and played saxophone with Archie Whitewater, and recorded an album on CHESS RECORDS;
  - Played saxophone on Malvina Reynolds' album, "Magical Songs for Children."
- Performed solo, as singer / song writer and guitarist, in East Coast clubs.

Organizing/Coordinating

- Co-produced theatrical dance shows for Bay Area audiences, as business partner in RAINBEAUZ PRODUCTIONS.
- Advised college faculty on modernizing curriculum and classroom technique using video and sound, as educational consultant for college curriculum committee.
- Organized and coordinated a European tour for the Howard Johnson sextet:
  ...Arranged travel ...Contracted for payment ...Interpreted in German.

EMPLOYMENT HISTORY

1992–present	***Bus Operator***	GOLDEN GATE BRIDGE DISTRICT, San Rafael, CA
1986–92	***Piano Tuner***	SELF-EMPLOYED, Forestville, CA
" "	***Co-Owner/Manager***	RAINBEAUZ Production, Forestville, CA
1980–86	***Musician/Music Teacher***	SELF-EMPLOYED, New York, NY
1977–79	***Educational Consultant***	CURRICULUM RESOURCES GROUP, Newton, MA

EDUCATION

B.A., Liberal Arts – SARAH LAWRENCE COLLEGE, Bronxville, NY

MARTHA AZCONA

1991 Greenhaven, Palo Alto, CA 94303
Home (415) 256-0121 • Work (415) 452-0057

Objective: Production position in special TV programming.

PROFILE

- 10 years' experience in TV news broadcasting.
- Graduate degree in documentary film making.
- Effectiveness in managing and directing others.
- Ability to develop a wide range of community contacts.
- Co-produced highly rated news magazine.

RELEVANT EXPERIENCE

Writing & Producing

- Co-produced daily syndicated news magazine, "PM":
  - Generated and researched story ideas;
  - Interviewed potential story sources;
  - Scouted filming locations;
  - Wrote scripts for 8-minute productions;
  - Directed production crew on location;
  - Supervised editing.
- Produced 13-minute show for United Way's "Project Match" on issues of housing for the elderly, at Osaka Productions.

Reporting & Interviewing

- As reporter for KTSM and KELP:
  - Developed contacts and stories, specializing in politics, health and education.
  - Developed and researched feature stories reflecting community interests.
  - Covered fast-breaking news stories.

Film and Videotape Editing

- Edited videotape for Osaka's 13-minute United Way promotional program.
- Edited videotape for "PM" news magazine shows.
- Filmed and edited stories for KGGM's local nightly news show.

Technical Expertise

- Developed working knowledge of:
  - Videotape editing equipment: Sony, JVC, RCA;
  - Videotape camera equipment: Hitachi, JVC, Sony, RCA, Ikigami;
  - Videotape recording equipment: Sony-BVU, JVC, RCA;
  - Film editing equipment: guillotine splicer, hot splicer, 16mm flatbed;
  - Film camera equipment: Bell & Howell, CP-16, Bolex, Beleau, Frezelini.

One way to highlight your technical skills.

BROADCAST HISTORY

1996–present	**Teaching Assistant**	STANFORD UNIVERSITY Communications Dept.
1995	**Creative Director**	OSAKA PRODUCTIONS – El Paso, TX
1993–94	**Producer**	KVIA-TV (for "PM" Magazine) – El Paso, TX
1990–92	**Photographer/Reporter**	KGGM-TV – Albuquerque NM
1989	**Interim News Director**	K102 RADIO – El Paso, TX
1985–88	**Reporter/Photographer**	KTSM-TV and RADIO – El Paso, TX
1982–85	**Reporter/Anchor**	KELP-TV – El Paso, TX

EDUCATION

M.A., Documentary Film Making, 1994 – Stanford University
B.S., Education (English; Journalism) University of Texas, El Paso
Professional Affiliation: National Association of Hispanic Journalists

Rebecca aims to get back into film production, which she loves. The functional format helps to play down her recent office work.

REBECCA NEWBURG

1901 Connecticut Avenue • South Orange, NJ 07079 • (201) 909-0006

Objective: Film production manager or assistant, in an agency or production house.

PROFILE

- 9 years' experience; working knowledge of all phases of production.
- Proven track record of producing and directing award-winning projects.
- Special talent for inspiring creative excellence on a shoestring.
- Resourceful; skilled in analyzing and solving problems.
- Practical grounding in the business and sales aspect of the industry.

MEDIA RELATED EXPERIENCE

Broadcast Production

- Produced and directed strip shows, weekly news magazine, live and taped specials, and difficult remote location shoots: – white water rafting – big game hunting – wilderness.
- Produced/directed investigative documentaries: i.e., youth detention, jail conditions, illegal aliens.
- Working knowledge of audio, 16mm, double system, studio, ENG, tape and film editing, mixing.

Film & Theater Directing

- Directed summer repertory theater at University of Idaho.
- Created and directed Children's Theater for Idaho Parks & Recreation Department.
- Directed 30-minute avant-garde dramatic feature film, "Cheesecake."
- Directed Radio Free Comedy, a professional New York-based improv group.

Producing & Production Management

- Developed sources of funding for film and tape projects, involving packaging, proposal writing, identifying, securing funding sources, capitalization, corporate sponsorship and trade-outs.
- Managed all aspects of a project: story development, budget, hiring, securing services.

Marketing, PR, Sales

- Developed relationship with corporations for sponsorship of Monarch Productions.
- Successfully secured national sponsors for Miss American Teenager Pageant of New Jersey.
- Built sponsorship, through successful cold calling, for a major cable magazine and cable TV.

WORK HISTORY

1995–present	***Executive Assistant***	BURKE & POWERS ADVERTISING, NYC
1993–94	***A.A. Corp. Communications***	CENTRAL CITY BANK, Pvt. Banking Div., NYC
1991–92	***Marketing Rep.***	NETWORK 4 ADVERTISING; So. Plainfield, NJ
1990–91	***Production Manager***	CBS CABLE DIVISION, NYC
1988–90	***Producer/Director***	IDAHO PUB. TELEVISION NETWORK; Boise, ID
1983–87	***Assoc. Producer/Director***	WJCT-TV, Jacksonville, FL

EDUCATION, TRAINING & AWARDS

M.P.P., Communications Policy – University of California, Berkeley 1983
B.A., Political Science, honors – University of California, Berkeley 1980

Film Production Awards:

Idaho State Broadcasters – Best Documentary, 1988,1989
Pacific Mountain Network – Best Short Subject, 1989

A "grip" is a person on a film production crew who assists the cameraman.

ROGER LANCASTER

9870 - 61st Street
Oakland, CA 94609
(510) 491-0098

OBJECTIVE **Entry level position as news grip with a TV station.**

PROFILE

- Cooperative teamworker in setting and achieving goals; personable, outgoing, approachable and communicative.
- Good eye for photography; experience with VCR camcorders.
- Strong interest in the field of TV reporting and camera work.
- Dependable, available, eager to learn, willing to work hard.

EXPERIENCE

Teamwork

- Served as waiter for exclusive party-yacht service, coordinating with co-workers on meticulously timed dinner cruises:
  - Met deadlines for finely detailed preparations;
  - Served four courses within strict time deadline, maintaining high quality and attention to detail.
- Participated in weekly meetings of a collective, taking into account everyone's opinions and priorities in working out problems.

Camera work

- Filmed 20 two-minute opinion interviews of Laney College students, operating a JVC camcorder.
- Additional experience handling a camcorder for personal use.

Project Support

- Assisted with order filling and customer services at Center for Local & Community Research:
  - Organized a wide range of data (including computerized info) to develop special interest mailings;
  - Answered phone inquiries, provided information, and referred customers to other sources of information in their field;
  - Filled mail order for books and pamphlets available from our company.
- Entered and sorted accounting data on a computer for a small accounting firm, successfully learning new procedures and following through without supervision.

WORK HISTORY

1996–present	**Waiter**	HORNBLOWER PARTY YACHTS, San Francisco
1995	**Dishwasher**	HORNBLOWER PARTY YACHTS, San Francisco
1994–95*	**Clerical Asst.**	CENTER FOR LOCAL & COMMUNITY RESEARCH, Berkeley (networking/info center for nonprofits)
1994*	**Accounting Asst.**	CARLENE COLE, Accountant, Berkeley

*part-time

EDUCATION Berkeley High School Graduate, 1995

ADRIENNE MENDOZA

2900 Ashby Place, #4
Berkeley, CA 94705
(510) 887-8779

Objective: Entry-level position in property management.

PROFILE

- Proven ability to quickly master new job skills.
- Resourceful; skilled in analyzing and solving problems.
- Easily develop rapport with clients and tradespeople.
- Experience in maintaining accurate records.
- Familiar with repair and construction terminology.

RELEVANT EXPERIENCE & SKILLS

Client Information • Negotiations

- Provided complex contractual information to prospective employers of university students under work-study program, interpreting federal and state employment regulations.
- Conducted field visits to campuses throughout the state, introducing computerized career information system to counselors and administrators:
  – Submitted detailed cost statements on lease of software and system installation;
  – Trained staff in computer terminology and use of the system.

Project Organizing • Problem Solving

- Successfully gained the confidence of resistant clients (employers, parents, students) in numerous situations through:
  – Consistent clarification of policies;
  – Listening to their point of view;
  – Negotiating workable compromises.
- Organized and maintained Job Placement Program for 1200+ students annually.
- Coordinated large campus Job Fair introducing students and employers:
  – Arranged for rooms and publicity;
  – Solicited participation of hundreds of employers.

Reporting • Record Keeping

- Documented extensive policy and procedure information, providing students with clear program guidelines and relieving staff of time-consuming verbal explanations.
- Oversaw reporting on student job placement program:
  – Calculated matching fund amounts;
  – Developed and maintained filing system documenting activity on 1000 student jobs;
  – Monitored job placement activity to assure compliance with program guidelines.

EMPLOYMENT HISTORY

1992–present	***Placement Interviewer***	FINANCIAL AID OFFICE, Hayward State University
1990–92	***Service Representative***	COMPUTER SYSTEMS INC., San Jose
1986–89	***Counselor***	GOODWILL INDUSTRIES, San Jose

EDUCATION

M.A., San Jose State University, 1986, Rehabilitation Counseling
B.S., San Jose State University, 1982, Child Development

Mark wants to apply his knowledge of the family real estate business to move out of his current job and into real estate.

MARK KILLORIN

219 Parker Street
Albany, CA 94706

(510) 425-0632

Objective: Trainee in Real Estate Property Management.

SUMMARY

- Experienced landlord and apartment complex manager.
- Licensed in real estate sales.
- Certificate in Real Estate from Diablo Valley CC.
- Lifelong exposure to family real estate business.
- 10 years' experience in retail sales.

RELATED EXPERIENCE

Property Management

- Managed 18-unit apartment complex:
  - Loan collections
  - Tenant complaints/requests
  - Yard work, maintenance
  - Screening and researching potential tenants
  - Monitoring parking
  - Appliances installation.

People Management

- Managed warehouse: verified accurate delivery; enforced strict receiving rules; prepared work assignments for night crews.
- Assistant Manager substitute:
  - Supervised 35 clerks
  - Resolved customer and employee disputes.

Real Estate

- Researched and purchased two income properties (a single-family dwelling and a duplex), both generating a profit.
- Developed expertise in all aspects of real estate financing.
- Assisted in accounting: loan recording and loan collections.
- Performed market analyses and square-foot analyses for property appraisals.

Sales

- Completed comprehensive course in Real Estate Sales at Diablo Valley College.
- Sold retail products and developed excellent customer rapport for 10 years, participating regularly in team competition for creative retail marketing displays

EMPLOYMENT HISTORY

1985–present	**Warehouse Manager**	P&S PRODUCE MARKET, Oakland
1993 summer	**Assistant Manager**	P&S PRODUCE MARKET, Oakland
1987–1988	**Office Assistant** part-time	E.J. KILLORIN Co., Real Estate Loans, Berkeley

EDUCATION

A.A., Business Administration – Diablo Valley College, Pleasant Hill
Real Estate Certificate Program, DVC
Real Estate Licensing Program, Anthony Schools, Oakland

Richard wants to apply his carpentry knowledge with his experience as a landlord, to a position in property management.

RICHARD C. FLORES

4849 Montana Blvd.
Pleasant Hill, CA 94523
(510) 692-1413

Objective: Position in Property Management with Wells Fargo.

SUMMARY OF QUALIFICATIONS

- Two years' experience in managing residential property as a landlord.
- Currently taking coursework in real estate practice.
- Extensive knowledge in all phases of new construction, remodeling and property maintenance.
- Trained by one of the area's most reputable construction firms.
- Outstanding ability to communicate with subcontractors in their language.

PROFESSIONAL EXPERIENCE

Construction Planning, Cost estimating

- Completely redesigned my own home, adding on 1000 square feet of improvements, complying with all building codes; acquiring permits; drawing up plans and scheduling subcontractors.
- Worked on construction of large commercial building and over 150 custom homes.
- Remodeled several residential properties, consulting with homeowners, selecting materials and estimating costs within their budgets.

Scheduling, Supervising

- Scheduled and monitored subcontractors (plumbers, electricians, sheet metal workers, roofers) for large scale custom home development project, in absence of foreman.
- Trained dozens of carpenter apprentices and new employees of park service.

Tenant Relations

- As landlord of a single-family unit:
  – Handled tenant complaints and arranged maintenance services;
  – Interviewed prospective tenants, made credit checks, drew up rental contracts.

WORK HISTORY

1988–present	Journeyman Carpenter	M.G. ROBERTSON CO. Walnut Creek, CA (residential home building)
1984–1987	Park Ranger	PARKS & RECREATION DEPT., Sacramento Co.
1982–1983	Assistant Manager	SAMBO'S RESTAURANT, Sacramento, CA
1980–1982	Park Aide (summers)	MT. DIABLO STATE PARK, Walnut Creek, CA

EDUCATION & TRAINING

B.S., Environmental Resources – California State University, Sacramento 1985
Fundamentals of Real Estate Practice – currently enrolled, Diablo Valley College
Carpenter Apprenticeship Program – completed 1993

AMY BUCHANNON

2700 Polk Street ♦ San Francisco, CA 94109 ♦ (415) 551-2866

Objective: Position as Graphic Designer's Assistant / Pasteup Artist

PROFILE

- ✦ Three years' experience in graphics design and layout.
- ✦ Working knowledge of pasteup techniques, mechanicals, ruling and assembly of type.
- ✦ Enthusiastic, energetic; excellent at working in a team setting to meet deadlines.
- ✦ Learn quickly; interpret information accurately.

GRAPHIC DESIGN EXPERIENCE

Graphic Design

- Created calligraphy for businesses and individuals, as free-lance calligrapher:
  - Designed wide variety of products: ...brochures ...poems/sayings for framing ...information signs ...certificates ...invitations ...fliers;
  - Designed and produced Book of Records for Pyramid Productions, involving calligraphy of 300 names of customers and officers, and ornamentation;
  - Selected character styles appropriate to the assignment;
  - Produced original and adapted designs for ornamentation and borders.
- Designed logos for community group, retail store, and professional therapists.
- Designed successful brochure for Narendra Bulow, bodywork therapist, choosing type, colors and materials consistent with desired image.
- Produced portfolio book illustrating basic production and design skills.

Pasteup

- Pasted up brochure of services for Surfaces Beauty Salon, involving calligraphy, photo reduction and pasteup.
- Pasted up mail order catalog for Emerald Priestess retail store.
- Pasted up many fliers for community organizations and small businesses.

Client Contact

- Consulted with graphics clients throughout production:
  - Assessed clients' needs and wishes;
  - Advised on creative and technical options for achieving desired image;
  - Showed comps for clients' approval.

EMPLOYMENT HISTORY

1995–present	Student	ACADEMY OF ART, San Francisco
" "	Graphic Artist	freelance
1994	Receptionist	ABOUT FACE & BODY salon, San Francisco
1994	Graphic Designer	EMERALD PRIESTESS, books & crystals, Berkeley
1992–93	Student	LANEY COLLEGE, Oakland

EDUCATION

Current enrolled – ACADEMY OF ART, San Francisco
– Figure Drawing – Silk Screening – Graphics Design: Materials, Tools & Techniques
LANEY COLLEGE: – Basic Design – Layout & Design – Lettering & Layout
FEATHER RIVER COLLEGE: Art History

BARBARA MONET
1219 Palo Alto Blvd.
Oakland, CA 94609
(510) 653-2211

OBJECTIVE

A graphic production position with a publisher, print shop, or publications department.

SUMMARY

- 4 years' experience as a graphics production artist.
- Familiar with the complete process of print production from first customer contact to bindery finishing.
- Able to organize and complete a complex project efficiently.
- Committed to high quality production and constantly sharpening skills.

RELEVANT SKILLS & EXPERIENCE

Production Organization

- Organized production of two 64-page yearbooks; did complete paste-up and assisted with typesetting and stripping.
- Produced first issue of a 32-page quarterly newsletter, including organizing of manuscript, proofreading, layout and paste-up.
- Coordinated production of 16-page monthly newspaper: arranged schedule with typesetter and printer, specified miscellaneous heads, managed last-minute production decisions.
- Acquired an overall understanding of small newspaper publishing by working in each production area:
  – paste-up artist; – office worker; – proofreader;
  – news editor / coordinator; – production coordinator.

Technical Skills

- Pasted-up a wide variety of camera-ready art, from basic business cards and letterheads to more demanding jobs such as posters, brochures and booklets involving color and screen tints.
- Prepared paste-ups in close collaboration with print shop stripper to assure highest quality and greatest efficiency.
- Operated reproduction camera for 20 years, preparing stats, negatives and halftones.

WORK HISTORY

1993–present	**Production Artist**	MIRO GRAPHICS CENTER, Berkeley
1992–93	**Paste-up Artist**	MARTINSEN LITHOGRAPHERS, Richmond
1990–93	**Production Artist/Editor**	WOMAN monthly newspaper, Alameda (range of editorial and production positions)
1989–90	**Asst. Manager/Cook**	RISTORANTE ITALIA, Oakland

EDUCATION

B.A., Humanities, UNIVERSITY OF CHICAGO
Graphic Arts, LANEY COLLEGE; Oakland, CA

GREGORY BYRON

2705 - 18th Street, #7
San Francisco, CA 94110
(415) 212-3551

Objective: Entry level position in multi-image production company.

EXPERIENCE

Print Media
- Operated graphic arts camera for Nowells Publications.
- Created concepts and layouts, specified type, and completed mechanicals for Northland's posters, brochures and books.
- Managed production department at Northland Press, a high quality book publishing company, supervising typography, stripping, printing, collating and binding.

Photography and Audio-Visual
- Shot still photographic documentation for Ecotechnics' oceanic research expedition.
- Developed a university-sponsored research project visually documenting eight innovative American communities, for visual design class at UC Berkeley.
- Completed introductory training in a computer graphics paint system, multi-image optical graphics, and multi-image programming.
- Currently working on free-lance multi-image productions.

Exhibits
- For Raychem Corporation, a high-tech Silicon Valley corporation:
  – Designed and built sculptures and exhibits using variety of media;
  – Fabricated and installed trade show exhibits;
  – Achieved expertise at commercial display silk screen technique.

EMPLOYMENT HISTORY

1994–present	**Exhibit builder**	Raychem Corporation, Menlo Park, CA
1994	**Freelance Graphic Designer**	Berkeley and San Francisco
1993	**Photographer, Electrician**	Institute of Ecotechnics, Santa Fe, NM
1990–92	**Project Manager, Fund-raiser**	Ecological Design Group, U.C. Berkeley
1989	**Production Manager**	Northland Press Book Pub., Flagstaff, AZ
1986–88	**Paste-up Artist, Cameraman**	Nowells Publications, Menlo Park, CA

EDUCATION & TRAINING

Technical Arts/Graphics - College of San Mateo, 1988
Sculpture - Northern Arizona University, 1989
Environmental/Visual Design - UC Berkeley, 1990–92
Computer Graphics - Academy of Art College, Fall 1989
Multi-Image Production - Artists in Print & Association for Multi-Image, 1996

Member: Northern California Chapter Association for Multi-Image

LYNNE CHARNEY

1990 Spruce Street • Berkeley, CA 94703 • (510) 245-8032

Objective: Design/production position in publications or advertising.

SUMMARY OF QUALIFICATIONS

- 10 years' professional experience, plus degree in fine arts.
- Sharp, innovative designer, in both B&W and color applications.
- Experienced in all aspects of graphics, from client needs assessment through camera-ready art.
- Effective in organizing large projects and coordinating with outside vendors.
- Communicate clearly and sensitively with clients and colleagues.

DESIGN & PRODUCTION EXPERIENCE

1994–present **Graphic Designer** Self-employed, Oakland

- Conferred with small business people and performing artists to clarify their graphic identity, target audience and written copy.
- Created variety of camera-ready art for business cards, letterheads, brochures, posters, ads, magazines.
- Designed highly effective logo and brochure for Cazadero Jazz Camp.
- Coordinated and produced 500-pg. catalog for Whole Earth Access Store, involving: • page layout • type spec'ing • pasteup • camera work • supervising an assistant.
- Designed and coordinated promotional materials for J.Miller Dance Theatre over a five-year period: • posters • announcements • fund-raising package • ads • stationery.

1995 **Graphics Technician/Drafter** CHEVRON USA, San Ramon

- Developed full color art work, with overlays, for scientific slide presentations.

1994–95 **Production Artist** FIFTH STREET DESIGN, Berkeley

- Illustrated high-tech manuals, advertisements, stationery, brochures, and geological maps, featuring both technical detail and artistic expression.

1991–94 **Fabric/Clothing Designer** LYNN CHARNEY HAND PRINTED SILKS, Oakland

- Created dynamic, commercially successful designs in unique colors for printed silk fabrics.

1983–94 **T-Shirt Designer** MARY & JOE'S SPORTING GOODS, Albany

- Designed custom T-shirts, applying my expertise in designing with type.

1980–82 **Pasteup Artist** THE MINES PRESS, New York

- Produced camera ready boards for several books and catalogs, including pasteup, page layout, camera-work, proofing, and type spec'ing.

EDUCATION

B.F.A. – California College of Arts & Crafts, Oakland
Studies in drawing, print making, design – New York University

– Portfolio available on request –

Vickie Wan

8888 Whitmore Street, Apt. 999 ◆ Oakland, CA 94611 ◆ (510) 987-6543

Objective: Position as **member of a design team** involving:
• Research/Problem Definition • Ideation/Initial Design • Prototyping

PROFILE

- Strong background in quick, thorough research resulting in workable ideas.
- Talent for incorporating the human element into design decisions.
- Proven ability to grasp a situation, adapt, and learn quickly.
- Able to meet deadlines and work with minimal supervision.
- Experience with multi-media as a communication tool.

TRAINING & EXPERIENCE

MULTI-MEDIA

Developed a design project to learn how multi-media could be used to enrich the learning experience:

- **Researched the field of multi-media** to determine exactly what it meant and how it could turn information into knowledge.
- **Learned HyperCard, Photoshop, and Addmotion** programs.
- **Developed user-scenarios through storyboards** to simulate how the program might be experienced.
- **Designed, assembled, and tested** a variety of screens, icons, and interactive elements, and arrived at a workable design.
- **Produced, evaluated, and finalized a working demonstration program,** refining the artwork, programming, and animation. *(Project available for demo.)*

EXHIBIT DESIGN

Created a scale-model exhibit design for a proposed "Visitor Center" and a related slide-presentation, to depict the experience of the exhibit.

- Learned to think critically, to appreciate the many aspects of exhibit design such as: user as audience; role of designer as social commentator; awareness of underlying concept and meaning.

PRODUCT DESIGN

Designed creative and playful products such as:

- Fish-shaped flashlight • Improved-design push-toy.
- Versatile, fully adjustable miner's lamp (involved significant research).

EDUCATION

B.A., Industrial Design, California College of Arts & Crafts, Oakland, 1994
4 years, Mechanical Engineering & Materials Science, UC Berkeley, 1986-90

EMPLOYMENT

1996–Present	Production Associate, Cogito Learning Media, San Francisco
1992–95	Part-time/summer office work for: ENSR Environmental Engineering, Xerox, Acura, Pennzoil, CitiBank (through Kelly Services)
1988–92	Production Assistant (part-time), Ten Speed Press, Berkeley

ANN VOORHEES
530 Battery Street
San Francisco, CA 94111
(415) 503-6365

Objective: Position as staff attorney with a government agency or professional organization.

SUMMARY

- 10 years' legal experience.
- Superior legal knowledge and skill, combined with a creative talent for using them to best advantage.
- Work well under pressure and enjoy challenging projects.
- Equally effective working independently or collaboratively.
- Committed to high ethical standards in the legal profession.

REPRESENTATIVE ACCOMPLISHMENTS & EXPERIENCE

Legal Expertise

- Worked as attorney with total responsibility for cases in a broad range of legal areas: ...Probate ...Estate planning ...Family law ...Civil litigation ...Property ...Tax ...Criminal law.
- Successfully completed all requirements for certification as Specialist in Family Law.
- Developed expertise in the field by keeping current on all cases and legislation in family law and related areas.
- Refined and strengthened legal skills through frequent continuing education programs.
- Serve regularly as Judge Pro Tem, Mountain Superior Court, Domestic Relations Mandatory Settlement Conferences.

Collaboration & Project Management

- Successfully managed a private law practice, training and supervising support staff and handling all accounting, billing and administration.
- Collaborated with court officials and interested professionals in writing and developing a videotape for use in the courts, to educate parents and facilitate resolution of custody and visitation disputes.
- Co-authored legislation, with other interested professionals, addressing training needs and job performance standards for city employees in sensitive positions.

EMPLOYMENT HISTORY

1991–present	***Attorney/sole practitioner***	Private practice, San Francisco, CA
1989–91	***Associate Attorney***	SMITH & CRANSTON, attys., SF, CA
1988–89	***Legal Researcher***	Self-employed independent contractor
1988	***Associate Attorney***	BERNARD FILMORE Esq., Palmdale, CA
1986–87	***Legal Intern/Law Clerk***	JUDGE MILTON SEBERT, Los Angeles;
	" "	Municipal Court Law & Motion Dept.;
	" "	JUDGE JOHN L. HARRIS, Culver City Municipal Court;
	" "	L.A. D.A. CONSUMER FRAUD UNIT;
	" "	LEGAL AID SOCIETY OF L.A.

– Continued –

ANN VOORHEES
Page two

EDUCATION & CERTIFICATION

J.D., UCLA School of Law, 1987 – Admitted to practice December 1987
M.A., History – University of California, Los Angeles 1982
B.A., History – University of Illinois, Chicago 1980
Certified Family Law Specialist, 1995

PROFESSIONAL AFFILIATIONS

- Executive Committee of Family Law Section, Los Angeles Bar Association
- Video Subcommittee, Los Angeles Bar Association
- Queens Bench
- Women in Criminal Justice
- California Women Lawyers

MICHAEL OLIVER

1888 Winthrop Avenue • Berkeley, CA 94709
(510) 303-6987 (home) • (510) 411-2627 (messages)

Objective: Position in law office as clerk, research assistant, or writer/editor.

SUMMARY

- Exceptional command of language.
- Dedicated professional attitude; mature and willing to work.
- Diligent, punctual, and extremely well organized.
- Extensive management experience.

RELEVANT EXPERIENCE

Legal-Related Experience
- Co-facilitated drug diversion therapy group (an alternative to incarceration) for Alameda County.
- Counseled clients with drug-related criminal histories, at residential treatment facility.
- Currently serving 3rd term as Administrative Committee Representative for student housing co-op, responsible for adjudication of complaints against the organization.

Research, Writing & Communication Skills
- Briefed numerous cases for Media Law and Criminal Law courses.
- Researched and wrote a manual for volunteer training at Ohlone Alcohol & Drug Services.
- Chaired meetings concerning house policy and finance for student co-op.
- Wrote job descriptions and annual reports, as Job Coordinator at Moyer House.

Management/Project Coordination
- Managed incoming accounts, coordinating successfully with both management and staff to increase production at large industrial laundry.
- Supervised work crews assigned to community jobs projects:
  ...trained personnel; ...developed work schedules; ...estimated and bid job costs.
- Oversaw kitchen, finance, and maintenance managers, as student co-op President.

EMPLOYMENT HISTORY

Current	Full-time student	UNIVERSITY of CALIFORNIA, BERKELEY
1996 summer	**Writer/Facilitator**	OHLONE ALCOHOL & DRUG SERVICES, Ohlone
1994–1995	**Job Coordinator**	MOYER HOUSE INC., Residential Treatment Facility Oakland
1993	**Resident/Counselor**	MOYER HOUSE INC, Oakland
1991–93	**Finance Manager**	MUMFORD HALL Student Residence, Berkeley
1992 summer	**Co-Manager**	ACME LAUNDRY, Chatham, MA
1987–91 summer	**Laborer**	ACME LAUNDRY, Chatham, MA

EDUCATION & LEGAL TRAINING

B.A., Drew University, Madison, NJ, 1981; Anthropology/Spanish Culture
B.A. in progress, Rhetoric Department – U.C. BERKELEY

Law-Related Courses:
• Criminal Law • Legal Philosophy • Media Law

– References available upon request –

RANDOLPH STROUGH

7900 Piedmont Avenue ■ Oakland, CA ■ 94618 (510) 412-4553

OBJECTIVE: Position as a paralegal.

PROFILE

- Strong analytical, writing and research skills.
- Solid grounding in litigation skills; completing paralegal certification program.
- Conscientious and thorough with detail.
- Equally effective working independently and in cooperation with others.

LITIGATION SKILLS

Writing

- Drafted interrogatories, declarations, and memoranda of law in handicap discrimination case.
- Drafted pleadings, discovery documents, client and demand letters, memoranda of law, and motions with points and authorities, in paralegal course work.

Research

- Performed research in handicap and sex discrimination cases:
  –motion to compel –application for preliminary relief in U.S. District Court;
  –response to motion to strike.
- Investigated public records in handicap discrimination case.
- Researched federal evidentiary questions in construction litigation case.
- Compiled case law on Sixth Amendment ineffectiveness of counsel in criminal appeals.

Trial Preparation

- Organized documents for trial and assisted in revising jury instructions in sex discrimination case.

Client Contact

- Interviewed Legal Services clients and represented them in administrative hearings.
- Interviewed immigration clients and assisted with INS forms, at Centro Legal La Raza.

EMPLOYMENT HISTORY

1995–present	**Student, paralegal**	SAN FRANCISCO STATE UNIVERSITY, Paralegal Certificate Program
1996	**Paralegal Intern**	EMPLOYMENT LAW CENTER, SF
1991–95	**Youth Counselor**	YOUTH HOMES, Walnut Creek (residence for disturbed adolescents)
1985–90	**Program Assistant**	UNIVERSITY OF CALIFORNIA, Berkeley
1983–84	**Paralegal**	BERKELEY NEIGHBORHOOD LEGAL SVCS.

EDUCATION, TRAINING & AFFILIATIONS

Paralegal certificate, San Francisco State University 1996 (GPA 3.6)
B.A., History – University of California, Berkeley 1984
Member, San Francisco Association of Legal Assistants 1996

Andrew is in his late 50's and has had no paid work experience in the past 25 years, outside of the family's antique firm. Emphasizing his recent internship and volunteer work increases his self-confidence for this major career change.

ANDREW THOMPSON

– Certified Legal Assistant –

1555 Powell St., #205 • Hampton CA 94000
(415) 793-4400

SUMMARY OF QUALIFICATIONS

- 1996 Graduate of Hampton College Paralegal Program.
- Completed internship with City Attorney's office, earning a reputation for high standards in teamwork and professionalism.
- Extremely thorough in working with details, and able to extract the most relevant points in legal documentation.

RELEVANT EXPERIENCE

PARALEGAL EXPERIENCE

As Paralegal in the Hampton City Attorney's office, working for attorneys John C. Brown, Zachariah Smith, and William Reardon:

- Summarized depositions, medical records, and employment records.
- Devised profiles of estimated earnings, losses, and medical expenditures.
- Prepared Notice of Entry of Judgment forms.
- Participated in 9-hour introduction to WESTLAW, a computer program for law offices.

LEGAL COURSEWORK (A.B.A. approved program)

...Legal Research & Writing	...Legal Concepts	...Contract Law
...Litigation I, II	...Real Estate Law	...Corporate Law I, II
...Probate	...Family Law	...Wills, Trusts, & Estate Planning

RESEARCH

- As Administrator/Librarian of Nursing School Library:
  – Researched available medical texts; made recommendations to department heads.
  – Classified books by specialized coding system.
- As Bibliographer for the University of California's Technical Services Dept.:
  – Conducted detailed cross-checking of library holdings to minimize costly duplication.

BUSINESS EXPERIENCE

As Secretary in family-owned antique furniture reproduction business in Seattle WA:

- Handled correspondence, phones, monthly inventories.
- Researched American and British sources of quality classic reproductions.
- Following sale of the business, co-managed the proceeds and assets of the firm.

WORK HISTORY

Fall 1996	**Paralegal Volunteer**	CITY ATTORNEY'S OFFICE, Hampton CA
May 1996	**Paralegal Intern**	CITY ATTORNEY'S OFFICE, Hampton CA
1980–present	**Assets Manager**	Managing and investing assets of Stanyan Ltd.
1974–80	**Partner/Secretary**	STANYAN LTD., antique reproductions firm
1968–73	**Librarian/Administrator**	NURSING LIBRARY, Clark Hospital
1967–68	**Bibliographer**	TECHNICAL SERVICES DEPT., University of California, Santa Cruz

EDUCATION & AFFILIATION

Paralegal Program – Hampton College, Hampton CA 1996
M.L.S., Library Science – University of California, Santa Cruz
B.A., Liberal Arts – University of Vermont, Burlington

Member, Hampton Association of Legal Assistants

RICHARD T. GRIFFON, JR.

5700 Riverview Street ● Albany, CA 94706 ● (510) 389-4111

OBJECTIVE: Part-time, freelance editing/proofreading.

SUMMARY

- Fast, accurate reader; extensive proofreading experience.
- Able to summarize written material concisely.
- Adept in organizing and integrating a number of documents into a coherent whole.
- Skilled in editing; able to write clear, precise prose.
- Experienced with major word processing programs.

RELEVANT EXPERIENCE

Editing & Proofreading

- Proofread articles written for economic trend forecasting newsletter.
- Served as Associate Editor of college newspaper; wrote articles, proofread articles submitted.
- Edited informative human interest magazine serving 5000 readers of Army Aviation Group.

Writing & Word Processing

- Wrote weekly reports detailing community action projects of Army battalions.
- Authored film reviews for small weekly newspaper.
- Wrote narration for local cable TV documentaries; wrote screenplay for three amateur films.
- Typed 130-page screenplay using Microsoft Word.

Project Organizing

- Supervised and coordinated the editorial direction and content, and specific writing assignments, of reporters for military publication.
- Organized the production of half-hour TV documentaries, making arrangements for acquiring equipment, and scheduling interviews and on-site video taping.

WORK HISTORY

current	Full-time student	GRADUATE THEOLOGICAL UNION – Berkeley, CA
1992–96	**Telephone Order Rep.**	ESPRIT clothing retailer/mail order – Portland, ME
1990–91	**Production Assistant and Writer**	PORTLAND HUMANITIES COMMITTEE (TV production house) – Portland, ME
1987–89	**Cameraman**	Free-lance work for Cable TV Co. – Portland, ME
1986–87	Travel/freelance writing	EUROPE
1985–86	**Proofreader**	RINFRET & CO., economic forecasters – NYC

EDUCATION

M.A. pending, Theology & the Arts – Graduate Theological Union, Berkeley
M.F.A., Film Production – New York University, NYC
B.A., English Literature, Baker University – Baldwin, Kansas

GEORGE AMUNDSEN
4414 Curtis Street, Berkeley, CA 94702
(510) 255-0457

Objective. Position as EDITORIAL ASSISTANT with Chevron Information Technology Co.

SUMMARY

- Successfully transformed "CHOBIZ" (Chevron HQ employees' magazine), improving its overall appearance and readability, while reducing costs.
- Skilled and experienced editor; dependable, industrious and creative.
- Adept in getting the most accomplished using the fewest resources.

PROFESSIONAL EXPERIENCE

Editing Publications

- Initiated a sharp new look to CHOBIZ, transforming it from a low interest, unorganized throwaway to a highly organized, visually stimulating monthly:
  - Created a system for maintaining accurate updated mailing list of 7500 names;
  - Coordinated layout and overall design in collaboration with design consultant;
  - Introduced the use of photo-typesetting, strengthening the overall appearance and readability of the paper, reducing costs, and simplifying proofreading.
- Co-edited in-house quarterly publication on Chevron Resources, THE PROSPECTOR, coordinating improved graphic design and reducing production costs.

Writing, Technical Writing & Copyediting

- Edited scores of articles in company publications for clarity, grammar, and style.
- Authored original articles & reports: editorials, feature articles, budget proposals.
- Wrote technical report documenting detailed start-up and management of a new business venture.
- Researched and wrote a report on computer graphics for cost and feasibility of in-house application.

Project Organization & Coordination

- Persuaded club leaders to write feature articles for company publication.
- Effectively negotiated the quick response and cooperation of a subcontractor to correct a recurrent mailing list problem, significantly saving time and money.

EMPLOYMENT HISTORY

1990–present	**Sr. Draftsman/Designer**	CHEVRON RESOURCES, San Rafael
1995	**Editor**, company publication	CHEVRON RESOURCES, San Rafael (concurrent with position above)
1989–90	**Laboratory Technician**	CHEVRON RESEARCH CO., Richmond
1987–89	**Purchasing Clerk; Lab. Asst.**	CHEVRON RESEARCH CO., Richmond
1982–87	**Graphic Artist**	FREELANCE; SELF-EMPLOYED

EDUCATION & TRAINING

B.A. Fine Art; Minor: Botany/Biology – Cal State Hayward, 1987

Additional Technical Skills And Education

Word processing, time-sharing, graphic art, computer graphics, computer programming
Courses: Computers for Management, Programming, Technical Writing

This college grad is still struggling to find her niche in the work world. She enjoys writing and research, and will use this resume to explore possibilities using these skills. She was advised to do some informational interviewing.

REBECCA BRIDGES

94 Silver Street ■ San Francisco, CA 94133 ■ (415) 909-3665

Objective: Position as Editorial Assistant/Researcher in publishing.

SUMMARY OF QUALIFICATIONS

- Keen perception for extracting important data.
- Proven successful in research writing.
- Strong skills in interviewing and developing rapport.
- Innovative in designing and carrying out projects.
- Highly motivated to achieve set goals.

PROFESSIONAL EXPERIENCE

Writing & Editing

- Authored Policy and Procedures Manual for the office of County Elections Manager:
  - Observed and documented manual office procedures and work flow in detail;
  - Interviewed employees to determine job description and time allocations for various tasks;
  - Analyzed data to establish procedural inefficiencies; made recommendations for improvement.
- Composed official correspondence to the constituents of State Assemblyman.

Researching/Analysis

- Researched and analyzed Mexico's macroeconomics policy in relation to foreign investment; presented synopsis of the 40-page report to graduate economic seminar.
- Conducted political research on constituent case needs for elected state official.
- Collected data regarding computer industry trends to advise and place computer professionals during job transition, at Bay Temp Service.

Project Management

- Managed a 6-month MBA recruiting project for Wells Fargo:
  - Screened resumes, contacted, selected qualified applicants for further interviews;
  - Developed a successful new system for processing qualified MBA candidates nationwide;
  - Presented written recommendations for project improvement.
- Founded and organized a professional development organization, Women in Political Science.

EMPLOYMENT HISTORY

Current	**Word Processor**	BAY TEMPORARY SERVICE, San Francisco
1996	Grad Student	San Francisco State University
1991–95	**Retail Sales**	CREIGHTON DEPT. STORE, San Francisco
1990	**Staffing Coordinator**	WELLS FARGO BANK, San Francisco
1989	**Technical Search Counselor**	HOWARD CORPORATION, San Jose
1988	**Procedures Analyst Intern**	FRESNO COUNTY ELECTIONS DEPT., Fresno

EDUCATION

B.A., Public Administration/Economics, 1988 – California State University
Graduate work in Economics – San Francisco State University

MARY MORIARTY

▪ Freelance Research ▪ Editing ▪ Collaborative Writing

901 Union Street, #70 • San Francisco, CA 94133
(415) 206 6162, call collect

Objective: Freelance work with books, articles and brochures.

- Copyediting • Creative/developmental editing • Acquisition editing
- Research and interviewing services for authors and writers
- Collaborative writing • Reading and evaluating book manuscripts

SUMMARY OF QUALIFICATIONS

- Strong commitment to excellence in the printed and published word.
- Demonstrated skill in developmental editing and copyediting.
- Over 10 years' experience coordinating library research services.
- Skill in assessing information needs for collaborative writing projects.
- Lifelong professional career evaluating and purchasing books for the public.

PROFESSIONAL EXPERIENCE

Copyediting, Creative & Developmental Editing

- Currently serving as series editor for books on health and nutrition; copyedited first book in the series.
- Developed and copyedited a book on color in interior design, working with author, publisher and illustrator to organize and refine the manuscript.
- Collaborated on writing and editing projects under contract with public relations firm:
  - personnel manual for a food manufacturer;
  - brochures explaining services of a medical professional group.

Research & Interviewing

- Researched and compiled subject bibliographies for three books currently under production.
- Interviewed restaurant owners and members of ethnic cultural associations for book on regional cooking and history.
- Managed reference services for the county library:
  - supervised staff;
  - interviewed reference patrons and formulated research strategies;
  - coordinated research networking services with San Joaquin Valley Information System.

Book Evaluation & Purchasing

- Evaluated manuscripts and produced reader's reports for two book publishers.
- Administered book acquisition budgets for city public library and for county library.
- Evaluated and selected books, using *Publishers' Weekly; Library Journal; Kirkus; Booklist;* book reviews in the *New York Times, Los Angeles Times, San Francisco Chronicle.*
- Wrote weekly book review column for community newspaper.
- Analyzed public reading tastes and consulted with publishers' representatives to stock and manage retail bookshop.

– Continued –

In the Employment History, Mary presents her various freelance jobs in a tidy package, and at the same time gets a lot of information across.

MARY MORIARITY
Page two

Subject Interests & Research Experience

- Self-help; psychology; sociology; women's studies.
- Medicine; health care; sexual relationships; marriage; parenting; aging.
- Parapsychology; erotica.
- Literature; fiction; mysteries; biography; letters; diaries.
- Bookselling; writing; publishing; book collecting.
- Food; cooking; restaurants; fashion; personal style.
- History; California; England.

EMPLOYMENT HISTORY

1996–present	***Editor***	FREELANCE, San Francisco & Central Valley:
	" "	...Martin-Moore Press, book publisher; and
	" "	...Dooley & Associates, public relations firm
	Manuscript Evaluator	Hitchcock Press, mystery publisher
1995	***Student***	Full-time studies at UC Berkeley Extension
1985–94	***Librarian III, Coordinator County Reference Services***	OTSEGO COUNTY LIBRARY, Oneonta, NY
1982–85	***Librarian II, and Asst. City Librarian***	OTSEGO COUNTY LIBRARY, Oneonta, NY " " "
1980–81	***Librarian I***	ONEONTA PUBLIC LIBRARY, Oneonta, NY
1978–79	***Library Assistant***	" " "
1974–78	***Owner/Manager***	VALLEY OAK BOOKSHOP, Oneonta, NY

EDUCATION & CERTIFICATION

B.A., Sociology – Stanford University, Stanford, CA
Publishing & Editing Certificate – UC Berkeley Extension

Specialized Training Courses:

- Publishing & Promoting the Cookbook, UCLA
- On-line Reference Service, San Jose State University
- Basic Data Processing, College of the Sequoias
- Interpersonal Skills for Library Management, UC Davis
- Collecting, Evaluating & Investing in Books, UC Davis

> The layout helps focus, with simple clarity, on advancement in Carolyn's most recent job.

Carolyn M. Clarke

115 Washington Blvd.
Sacramento, CA 95820
(916) 605-4777

OBJECTIVE: Position as editorial assistant in book publishing.

SUMMARY

- Lifetime interest in books; willing to learn all aspects of publishing.
- Substantial experience in writing, rewriting, and editing.
- Well organized and thorough in completing complex projects.
- Committed to high quality production and attentive to details.

PROFESSIONAL EXPERIENCE

Project Coordination

- Developed and implemented in-service training for savings and loan employees:
  – Set training schedules; – Selected trainees; – Determined curricula needs;
  – Wrote training materials (quizzes, case studies, work sheets);
  – Wrote presentations; – Conducted training sessions.
- Administered and coordinated daily operations and long-term planning of retirement plan program involving 160,000 accounts and $560 million in deposits:
  – Hired, trained, and supervised staff of six;
  – Consulted with marketing, legal, and computer departments on proposed policy and procedure changes;
  – Developed promotional strategy to maintain competitive edge in customer services;
  – Researched federal legislation to assure continued compliance with new laws.

Editing and Publishing

- Developed glossary of terms, quick reference guides and employee workbooks to be used by 200 branch offices of City Savings and Loan.
- Wrote, edited, and published weekly procedural bulletins for company employees.
- Designed and/or redesigned content and layout of retirement plan forms and booklets.

EMPLOYMENT HISTORY

1994–present	***Retirement Plans Administrator***	CITY SAVINGS AND LOAN, Sacramento, CA Retirement Plans Department
1993	***Special Projects Coordinator***	" " "
1992	***Research Assistant***	" " "
1991	***Customer Services Specialist***	" " "
1986–90	***Salesperson***	COFFEE EXPERIENCE, Sacramento, CA MULLIGAN'S PHARMACY, Davis, CA
1983–85	***Secretary***	GENEVA ENGINEERING, Sacramento, CA
	Secretary	DR. W. MARTIN, Medical Office, Sacramento, CA

EDUCATION

B.A., Political Science, 1990 – University of California, Davis, CA
Knowledge of Spanish, French, Norwegian, Swedish
Course work in: English literature and composition; Scandinavian literature.

LINDA DURKEE

1219 Shafter Ave.
Oakland, CA 94609
(510) 694-3336

Objective: Entry position in Real Estate Appraisal.

SUMMARY

- Currently enrolled in Real Estate Appraisal course; committed to developing expertise in real estate appraisal.
- Direct experience assessing the condition of a building.
- Skill in researching and comparing market information.
- Talent for mathematical computation and data organization.

RELEVANT EXPERIENCE

Real Estate Knowledge

- Purchased and managed a triplex rental unit:
  - Conducted a comparative search of properties in Oakland and Berkeley;
  - Completed course for owners on home repair;
  - Handled minor plumbing and electrical problems;
  - Researched construction, foundations, and earthquake precautions.
- Studied Real Estate Appraisal and Principles of Real Estate.

Research • Data Organization • Computation

- Maintained detailed financial and material records for jewelry business, documenting:
  ...gross income ...expenses ...inventory ...sources and comparative costs of materials.
- Recorded rental operating costs and computed annual expenditures to assure profitability.
- Researched and contracted to purchase skilled labor, building appliances and materials, as owner/manager of a rental triplex.
- Researched commercial sources of gems, metals and related products from various manufacturers to determine best combination of price and quality.

EMPLOYMENT HISTORY

1986–present	***Jeweler/Mgr./Mfg./Sales***	LINDA DURKEE JEWELRY DESIGN, Bay Area
1991–present	***Property Manager***, part-time	Building Owner & Manager, self-employed
1978–85	***Sales/Administrative Asst.***	DAYTON POOL & SUPPLY CO., Dayton, OH
1992–95	***Counselor***	GOODWILL, San Jose

EDUCATION & TRAINING

Psychology major, OHIO UNIVERSITY – Athens, OH

Real Estate Related Courses

- Principles of Real Estate • Accounting
- Real Estate Appraisal, currently enrolled

Margaret got advice from a pro in her chosen field about what to put on her resume. Note that she itemizes the tools she can work with.

MARGARET LESTER
1919 Richardson Lane
Hayward, CA 94541
(510) 256-1990

Objective: Position as real estate appraiser trainee.

PROFILE

- Property owner with personal experience in real estate and home maintenance.
- Sharp, quick learner; willing to get involved.
- Effective working alone and as a cooperative team worker.
- Strength in analyzing, researching, organizing, and problem solving.
- Reliable and hard working; thorough in completing projects.

RELEVANT EXPERIENCE

Real Estate

- Researched real estate market areas extensively, on financial and desirability criteria.
- Performed home maintenance, including: ...wall papering ...painting ...wall patching ...insulating ...landscaping.
  Handy with power drills, skill saw, table saw, T-square, sander, buffer, auto, pick-up, van.
- Purchased two homes and sold one.

Math & Finance

- Computed interest on savings accounts; trained other staff in computations.
- Calculated daily cash receipts and prepared deposits for bookstore and bank.
- Handled a wide range of customer banking products and services: –savings accounts –IRAs –CDs –account loans –loan insurance –financial inquiries.
- Traced detailed history of client bank records to resolve customer service issues.

Business Writing

- Wrote bank correspondence advising and counseling bank customers on new accounts and services available.
- Completed studies in "Effective Business Writing" and "Communication Skills for Business."

EMPLOYMENT HISTORY

1988–present **Branch Manager** UNIVERSAL SAVINGS BANK, Oakland, CA
1983–1988 **Elementary Teacher** OAKLAND PUBLIC SCHOOLS
1980–1982 **Acting Asst. Mgr./Sales Clerk** GOLDEN STATE BOOK CO., Oakland, CA

EDUCATION

M.A., Education, 1986 – HOLY NAMES COLLEGE, Oakland, CA
Standard Teaching Credential, Elementary, 1982 – HOLY NAMES COLLEGE
B.A., Sociology, 1982 – HOLY NAMES COLLEGE

Aiming for a career change, Dennis pulls together his recent coursework and skills that support an entry level position in the new field. Overlapping periods in work history are made clear.

DENNIS BRINKLEY
987 Lakemead Way
Redwood City, CA 94062
(415) 901-6001 work
(415) 511-6887 home

Objective: Trainee in real estate appraisal.

SUMMARY

- Self-motivated, mature, focused and ambitious.
- Successful in mastering new skills through hands-on experience.
- Proven ability to reach accurate, objective conclusions involving a great deal of data and variables.
- Skill in identifying the real goal, and finding ways to achieve it within available time, resources and conditions.

EXPERIENCE & ACCOMPLISHMENTS

KNOWLEDGE OF REAL ESTATE, FINANCE & ECONOMICS

- Completed AIREA course in Principles of Real Estate Appraisal.
- Mastered real estate laws and procedures, earning California licensure in real estate sales.
- Authored an abstract on the impact of banking and money on macroeconomics; predicted the necessity for American companies to merge with foreign competitors to survive.

ANALYSIS OF VALUE & FEASIBILITY

- Appraised the feasibility and cost effectiveness of company's proposed expenditures:
  – plant improvements – major purchases of equipment – contracts
  – property and equipment leases – new facilities – equipment and service upgrades.
- Transformed a chaotic book warehouse into a highly efficient and cost-effective operation:
  – Assessed the current conditions, traffic flow, and turnaround time for processing goods;
  – Developed theories to best utilize available space and manpower to increase production;
  – Implemented and modified these new ideas until the desired results were achieved;
  – Results: saved thousands of dollars in labor and freight costs; increased customer satisfaction; improved quality of the product; met production deadlines and quotas.

PROJECT PLANNING/MANAGEMENT

- Successfully managed the start-up of a local TV station:
  – Assessed needs of community through public and private meetings;
  – Found local talent and trained volunteers for all aspects of TV production;
  – Supervised production of final programming.
- Developed scholarly journal from initial concept to actual production and distribution, learning all aspects of magazine production in the process.

EMPLOYMENT HISTORY

1995–present	***Comptroller***	ACE BOOK DISTRIBUTORS, Brattleboro, VT
1994–95	***Operations Manager***	ACE BOOK DISTRIBUTORS, Brattleboro, VT
1987–88	***TV Production Manager***	BRATTLEBORO COMMUNITY TV (concurrent with below)
1985–93	***Editor/Writer***	ZAHRA PRESS, London, U.K.

EDUCATION & TRAINING

B.A., Social Sciences – UNIVERSITY OF CALIFORNIA, BERKELEY
Real Estate Sales License, California
Course in Real Estate Appraisal Principles, AIREA

SUSAN G. HOLMES

9000 Le Conte Avenue, Apt. 78
Berkeley, CA 94709
(510) 244-6161

Objective: **Position as Pastoral Minister, assisting in care of parish family.**

SUMMARY OF QUALIFICATIONS

- M.Div./M.A., May 1996
- Experience in Pastoral Counseling and Worship Planning.
- Effectiveness in recruiting, motivating, placing, training and recognition of parish volunteers.
- Strong commitment to adult Catholic education, and success in teaching adults.
- Ability to deal with diverse individuals and groups, including minorities.
- Strong background in Carmelite and Ignatian spirituality, and interest in providing for the spiritual needs of individuals.

RELEVANT EXPERIENCE

Counseling & Pastoral Service

- Provided general pastoral counseling, as:
  - hospital chaplain intern;
  - interviewer for RCIA Program responsible for recommending advancement of Catechumens;
  - member of House of Prayer team; provided consultation as needed;
  - minister to bedridden, convalescent hospital;
  - member in group process and group spiritual direction in parish prayer group.
- Counseled for Suicide Prevention hotline of Alameda County.
- Served as Lay Minister of the Eucharist for: eucharistic liturgies; prayer group; bedridden.
- Interviewed in parish census and survey to assess needs in Parish Council effort.

Religious Education

- Planned one year RCIA Program as member of Core Team: set calendar, evaluated sessions, prepared liturgies.
- Directed children's religious education programs:
  - Made teacher development a priority;
  - Coordinated sacrament classes;
  - Selected new texts;
  - Relocated teaching facility;
  - Monitored class content and teaching.
- Designed/presented classes for CCD parents and teachers, on sacraments, teaching methods.

Worship

- Planned eucharistic liturgies: ...Sunday parish liturgy, for two years;
  ...Children's liturgies, on several special occasions.
- Conducted Communion Services at Convalescent Hospital and in parish as Eucharistic Minister.
- Planned evening of recollection for CCD teachers.

– Continued –

SUSAN G. HOLMES
Page two

RELEVANT EXPERIENCE, Continued

Administration

- Served on Parish Council, both as college student rep and at-large rep.
- Managed reservations calendar and retreatants' quarters for House of Prayer.
- Coordinated two Religious Education programs, including:
  - Managing and training 40 volunteer teachers and aides; active recruitment of minorities;
  - Liaison among three Chaplains and a Sunday School Superintendent;
  - Inventorying materials and light budgeting;
  - Establishing on-site audiovisual and reference library.

MINISTERIAL HISTORY

1995–96	**Chaplain Intern**	Samuel Merritt Hospital, Oakland
1994–95	**Director of Religious Education**	Naval Station Chapel, Alameda
1993–94	**Core Team Member, RCIA Program**	Saint David's Parish, Richmond
1992–93	**Crisis Counselor**	Suicide Prevention & Crisis Intervention, Alameda County
1990–91	**Acting Guest Master**	Carmelite House of Prayer, Oakville
1988–89	**Lay Minister of the Eucharist**	Holy Spirit Parish & Elmwood Convalescent Hospital, Berkeley
1987–89	**Core Member, Prayer Group**	Holy Spirit Parish, Berkeley
1981–84	**Member & Chair, Liturgy Committee**	" " "
1981–82	**Member, Parish Council**	" " "

EDUCATION

1992–96 M.A. in Theology;
Dominican School of Philosophy & Theology (at Graduate Theological Union) – Berkeley, CA

1992–96 M.Div., Dominican School of Philosophy & Theology (at GTU) – Berkeley, CA
(strong emphasis in scripture, moral theology and social justice, New Code of Canon Law, sacramental theology, liturgy and homiletics)

1979–81, 1983–85 B.A., Ancient History & Archaeology (honors) – UC Berkeley

COMMITTEES

1995–96 Member, Planning Committee;
Workshop on Sexual Abuse for Ministers

1993–94 Member, Lay Student Committee;
Dominican School of Philosophy & Theology (at GTU)

AWARDS

1995 Mickey Award (Scholarship – Essay Competition) Graduate Theological Union, Berkeley

– References available upon request –

SISTER MARY JONES

234 Hampton Road
Carlton, MA 12345
504-555-4444

OBJECTIVE: Pastoral Associate.

SUMMARY OF QUALIFICATIONS

- Possess desire and ability to reach out and serve others personally.
- Have learned how to *listen* to people — to hear what is said and unsaid.
- Accept and appreciate people for who and what they are — am enriched by each association.
- Have worked with people in many settings: as confidante and counselor, workshop instructor, RCIA sponsor, adult literacy tutor, spiritual friend.
- Hard-working, energetic, and optimistic.
- Master of Divinity degree.

PASTORAL ACTIVITIES AND EXPERIENCE

- Have served as **Eucharistic Minister and Lector** at St. Joseph's Church, Carlton, MA, since 1984. Also take Eucharist to the **sick and homebound.** Among other activities at St. Joseph's:
  - Served as **Sponsor** and in other roles with **RCIA program,** 1984-92.
  - Was part of a team facilitating **Landings Program,** providing **information and support to inactive Catholics** considering possible return to the Church, 1993-94.
- Served as **Chaplain** to St. Anne's Villa community in absence of Msgr. Riley several weeks yearly, 1991-95. **Led congregation** in Liturgy of the Word and Communion.
- Completed 12-week **Clinical Pastoral Education** placement, Central Hospital, New Orleans, LA, 1982. Earned one semester of credit toward completion of Master of Divinity at Harvard.
  - Served as **Chaplain on medical/surgery floor** (24-hour on-call assignment, Saturday 8 a.m. - Sunday 8 a.m.). Provided **support and assistance to seriously ill and injured patients** and their families. Conducted 1 a.m. **Sunday worship service.**
  - Performed Wednesday morning **Chapel services** on alternating schedule.
- **Tutored for Metro Adult Literacy Foundation** since 1985. Also **trained other volunteers** to tutor illiterate adults.
- Served as **Chaplain for separated-and-divorced group** in Newtown, ME, 1978-81.

– Continued –

RELEVANT EMPLOYMENT HISTORY

- Assisted separated and divorced individuals involved in nullity process, as **Advocate/Assessor at Diocesan Tribunal,** Carlton. (1984-Present)
  - **Conducted in-depth personal interviews** with petitioners, spouses, and witnesses (family, friends) to gather detailed history of the marriage.
  - **Led them through sensitive marriage issues** — backgrounds of couple, how they met, the courtship, marriage, and presence or absence of elements constituting a sacramental marriage.
  - Gained valuable experience in **listening** and in **helping others to express themselves** concerning personal and spiritual matters.
- **Prepared and delivered informational presentations** at separated-and-divorced conferences for those considering nullity process. Addressed both group and individual questions. (Since 1984)
- Participated in **marriage preparation for engaged couples** since 1984, especially those entering subsequent marriages. Member of team conducting weekend retreats at St. Martin's Hall.
  - **Led workshops** concerning issues characteristic of subsequent marriages, such as combined families.
  - Was part of a team that **developed written guidelines** for couples entering subsequent marriages.
- As **Faculty Librarian**, St. Luke's Provincial Seminary, Newtown, ME, was often **sought out as confidante** by seminarians and lay students concerning personal and spiritual issues. (1978-1981)
- Served as **spiritual friend to inmates**, as **Librarian**, Joliet Correctional Center, Joliet, IL. Supervised inmate library clerks. (1971-78)

EDUCATION

- Master of Divinity degree, 1984
  Harvard University, Cambridge, MA
  (Qualified for financial assistance based on scholarship.)
- Master of Arts degree, Library Science, 1971
  Rosary College, Bluefield, MI
- Bachelor of Arts degree, English and History, 1967
  DePaul University, Chicago, IL

Continuing Education includes:

- *Stillpoint,* course in spiritual direction, 1993-present (ongoing)
  (Rev. Sue Smith, St. John's Episcopal Church, Carlton, MA)
- Crisis Intervention Center, Carlton, winter 1995
  50 hours of training as Counselor
- Completed a number of Tribunal training workshops.
- Thirty-day Closed Retreat.

AFFILIATIONS

Religious of the Institute of the Blessed Virgin Mary

– Resume written by Carolyn Schmitz –

JO ANNE BURGESS

1266 San Gabrielle Ave. • Oakland, CA 94619 • (510) 282-6900

Objective: Position as Union Representative / Business Agent; or Labor Organization Administrator.

SUMMARY

- 12 years' experience in union organizing, contract negotiations and grievance handling.
- Committed to workers and their needs.
- Exceptional ability both to communicate with members, and to articulate the union's demands to management.
- Demonstrated effectiveness in sizing up a situation and getting the job done.

PROFESSIONAL EXPERIENCE

Contract Negotiations

- Effectively negotiated scores of contracts in:
  - Transportation companies;
  - Financial institutions;
  - Dairies, bakeries and food industries;
  - Hospitals and health maintenance organizations.
- Handled all aspects of negotiations:
  - Ascertained the needs of the membership through meetings and surveys;
  - Formulated proposals designed to resolve identified problems, needs and demands;
  - Served as chief negotiator working with rank and file committees;
  - Drafted and finalized contract language.

Grievance Handling & Contract Enforcement

- Represented hundreds of grievants covered by diverse contracts.
- Prepared and presented grievances for arbitration.
- Monitored contract agreements through contact with shop stewards and members.
- Represented members in dealings with state and federal agencies.

Organizing Workers

- Directed, and strategized successful organizing campaigns with rank-and-file committees.
- Prepared programs and presentations; spoke at meetings, rallies and demonstrations.
- Authored and edited many organizing leaflets and brochures.

Administration / Management

- Coordinated a small union office, implementing the directives of union officers, executive board, committees and members: collected dues; responded to inquiries; wrote, designed, and distributed leaflets; advised volunteers; processed mass mailings.
- Supervised and directed the field work of four union representatives during absence of the Senior Union Representative.
- Developed strategies for the union's participation in strikes and strike support activities, in conjunction with other labor organizations.
- Prepared press releases and handled media relations to publicize union activities.

WORK HISTORY

1986–present	**Union Representative**	OFFICE AND PROFESSIONAL EMPLOYEES UNION, LOCAL 29, Emeryville, CA
1982–86	**Union Organizer and Office Secretary**	AMERICAN FEDERATION OF STATE, COUNTY AND MUNICIPAL EMPLOYEES, LOCAL 1695, UC Berkeley

BRADLEY FRENCH

P.O. Box 900 • San Rafael, CA 94915 • (415) 459-2008

Job objective: Writer / Photographer / Editorial Assistant position with a newspaper, magazine, PR firm or book publisher.

SUMMARY

- Successfully published writer, editor and photographer.
- Enthusiastic and committed; go-getter who doesn't quit until the job is done right.
- Effective problem solver; thorough researcher.
- Well organized and focused in coordinating projects.

PROFESSIONAL EXPERIENCE

Writing & Editing

- Wrote feature articles for national magazines, including *ALL ABOUT BEER*, as Northern California field editor and photojournalist.
- Wrote KQED Beer Festival Guide for *San Francisco FOCUS* magazine's July 1986 issue.
- Created and published a local specialty newsletter for home brewers / collectors:
  - Pub and book reviews;
  - Local events and openings;
  - New products and recipes.
- Selected, proofread and copy edited manuscripts for Stonehenge Books.

Photography

- Published photographs for:
  - Magazines: produced product shots, location and personality photos, for *ALL ABOUT BEER* Magazine;
  - Newspapers: photo series on author Ann Rice in *FICTION MONTHLY*;
  - Book: "The Elitch Gardens Story," published by Rocky Mt. Writers Guild, Boulder, CO;
  - Slide show: Sierra Club's national slide slow, "The Ultimate Environmental Issue;"
  - Photographed models for glamour and fashion for Vannoy Talent Agency.

Public Relations

- Implemented successful marketing campaigns for Jack London books at Star Rover.
- Headed promotion and PR department for Stonehenge Books, Denver book publisher:
  - Arranged media interviews for new authors; initiated a new weekly radio program featuring interviews with Stonehenge authors;
  - Organized book signing publicity events in area book stores;
  - Wrote press releases and submitted review copies to book columnists;
  - Developed mail order book promotion directed toward special interest groups.

FREELANCE WORK HISTORY

Current	***Writer/Photographer***	FREELANCE; most recent assignments for: *NETWORK MARKETING, PRACTICAL WINERY, AMATEUR BREWER*
1992–96	***Field Editor,*** *No. Cal.*	*ALL ABOUT BEER* MAGAZINE, Anaheim, CA
1993–95	***Book Sales Rep.*** *(concurrent with above)*	STAR ROVER HOUSE, book publisher, Oakland
1991–92	***Book Sales Coordinator***	DETERMINED PRODUCTIONS, SF, books / gifts mfg.
1989–90	***Editing/Mktg. Asst.***	STONEHENGE BOOKS, book publisher, Denver, CO

EDUCATION

B.A., Anthropology – Arizona State University; Tempe, AZ
KIIS Broadcasting Workshop, Hollywood, CA • FCC 3rd Class License

LOREN GREENE

1415 Pine Avenue
Richmond, CA 94805
(510) 776-3009

Objective: Position as Summarizer for Barron's Legal Services.

SUMMARY OF QUALIFICATIONS

- Two years' experience as legal aid/intern.
- Effective at translating complex technical information into easily understood language.
- Enthusiastic and dedicated professional with strong analytical skills.
- Ability to work independently and as a cooperative team member.

PROFESSIONAL EXPERIENCE

Technical Writing

- Prepared written reports, correspondence and policy papers for PG&E, articulating complex issues and recommending positions.
- Wrote database users' guide for PG&E employees.
- Produced summary reports on California hazardous waste litigation.
- Documented 20th-century U.S. land use and energy policy strategies.

Analysis & Evaluation

- Researched utility rate-setting procedures for Wisconsin Citizens' Board.
- Developed and implemented new policies at PG&E to address issues in all areas of small hydropower development.
- Formulated policy recommendations, incorporating information gathered from various departments at PG&E.
- Advised private developers on policies affecting their proposed alternative energy projects.
- Designed database procedures, accommodating information needs of PG&E employees.

Project Management

- Served as project liaison between developers and the PG&E system.
- Implemented regulatory requirements, monitoring hydro projects and documenting developer compliance.
- Participated in planning and implementing of prairie restoration project.
- Supervised staff of 10 field canvassers in community outreach techniques.

EMPLOYMENT HISTORY

1996–present	**Resource Analyst**	PACIFIC GAS & ELECTRIC, Walnut Creek, CA
1996–present	**Legal Aid**	LEGAL ASSISTANT SERVICES, Orinda, CA
1994–95	**Project Assistant**	UNIV. OF WISCONSIN ARBORETUM, Madison, WI
1992	**Field Manager**	CITIZENS' UTILITY BOARD, Madison, WI
1991	**Legal Intern**	STATE UNIVERSITY OF NEW YORK, Potsdam
1990	**Library Assistant**	STATE UNIVERSITY OF NEW YORK, Potsdam

EDUCATION

M.A., Political Science – UNIVERSITY OF WISCONSIN, Madison, WI 1996
Area specialty: Environmental Policy

Michael, a recent graduate, is hoping to use his social science degree in his work. He'll need to do some research and informational interviewing and then come back to shape his resume.

MICHAEL WONG
1914 - 12th Avenue
San Francisco, CA 94122
(415) 688-0900

Objective: Position as Community or Governmental Relations Representative.

SUMMARY OF QUALIFICATIONS

- Over 8 years' experience in community relations work.
- Graduate degree with emphasis in public administration.
- Effective and persuasive with all segments of the community.
- Skilled and thorough in analyzing problem situations and finding creative solutions.

PROFESSIONAL EXPERIENCE

Community Relations

- Advised and offered technical assistance to United Way agencies on fund-raising, budgeting, program planning.
- Served, by appointment, on Housing Committee charged with analyzing local housing crisis and providing follow-up recommendations.
- Coordinated successful special event and direct mail fund-raising for local city council members.
- Developed and implemented a successful marketing campaign (where others had repeatedly failed), introducing new students to local business services.
- Acted as liaison between student government and campus community, coordinating meeting times, places, and agendas.

Management & Administration

- Established policy for Humboldt Housing Action Project and University Activities Center; evaluated overall program effectiveness and employee performance.
- Trained and supervised volunteers and paid employees in counseling and fund-raising.
- Issued general press releases and public service announcements for voter registration drive.
- Chaired meetings of up to 100 people.

EMPLOYMENT HISTORY

1993–present	**Fund-raiser**	ROGERS-WILLIAMSON PARENT COUNCIL, Sierra, CA
1992–93	**Planning Intern**	ARCATA PLANNING DEPARTMENT, Arcata, CA
1992–93	**Sales Manager**	ASSOCIATED STUDENTS, Humboldt State Univ., Arcata, CA
1992	**Retail Sales**	REDWOODS UNITED, Eureka, CA
1989–90	**Researcher**	NATIONAL INSTITUTES OF HEALTH, Eureka, CA

EDUCATION

M.A. Social Science, Humboldt State University, Arcata, CA
Emphasis in Public Administration (thesis in progress)
B.A. Social Science, Humboldt State University, Arcata, CA

COMMUNITY & VOLUNTEER EXPERIENCE

- United Way, Bay Area and Humboldt County – Allocations Team Member.
- Board of Directors: Humboldt Housing Action Project; University Activities Center.
- Chairperson: Student Legislative Council, Humboldt State University.
- Member of Arcata Housing Task Force (appointed by City Council).
- Coordinator, voter registration drives.

RICHARD JENNINGS

8009 Mountain Blvd. • Oakland, CA 94602 • (510) 111-6887

Objective: Entry level position as fire fighter with a municipal fire department.

SUMMARY OF QUALIFICATIONS

- Able to function at top performance throughout a 24-hour shift.
- Proven ability to respond immediately and confidently in emergencies.
- 3 years' experience as a paramedic; 5 years as an emergency medical tech.
- Committed professional, constantly upgrading training.
- Excellent relations with the public and the community.
- Mature lifestyle, compatible with emergency work.

RELEVANT EXPERIENCE

Crisis Evaluation & Response

- Effectively evaluated thousands of emergencies, for example:
  – auto accidents requiring treatment before transport;
  – heart attacks (80 Code Blues this year, with approximately 20% save rate);
  – family interventions (treating family members with respect and sensitivity).
- Adapted immediately to changing circumstances in medical emergencies, setting priorities and constantly reevaluating them.

Community Relations

- Served as liaison between paramedics, hospitals and fire department:
  – Mediated communication problems among professionals;
  – Provided monthly training updates for the firemen I worked with;
  – Provided follow-up medical information on cases jointly handled.
- Educated the public on the role of emergency medical services, through demonstrations, lectures and small group talks.
- Taught public CPR classes; served as volunteer medic for public events; trained nurses and hospital employees in trauma skills at Eden Hospital.

Medical Teamwork

- Highly skilled in all basic medical emergency techniques:
  ...taking blood pressure and pulses ...wound management ...CPR
  ...splinting fractures ...applying oxygen ...patient assessment.
- Led ambulance team, as Crew Chief for Bay Medical Services.

Quality Assurance

- Evaluated new 911 provider paramedics on the job, as Alameda County paramedic evaluator, assuring they understand the protocols and procedures, and perform to county safety standards.
- Wrote training manual for new medics employed by Bay Medical Services.
- Selected for and served on Paramedic Peer Review Committee at Eden Hospital, monitoring paramedic response to improve treatment of patients.
- Trained new paramedics in the field portion of their state-required training time, focusing on communication and decision-making skills in medical emergencies.

– Continued –

Richard is considering a career change from EMT (see page 154) to fire fighting. He'll use this version of his resume as a first draft for informational interviewing. When he learns more about a fire fighter's job, he may modify his resume.

RICHARD JENNINGS
Page two

EMPLOYMENT HISTORY

1995–present	**Paramedic Crew Chief**	BAY MEDICAL SERVICES, Berkeley / Oakland
1994	**Paramedic**	KING AMERICAN AMBULANCE, San Francisco
1992–94	**EMT-1A**	ALLIED AMBULANCE, Oakland
1987–91	**Warehouseman/Driver**	GOOD GUYS; SAUSALITO DESIGN; PACIFIC FLOORING retail businesses, SF Bay Area
1984–86	**Day Care Driver**	EASTER SEAL SOCIETY, San Francisco

EDUCATION & SPECIALIZED TRAINING

B.A., Sociology – SAN FRANCISCO STATE UNIVERSITY
A.A., Criminology – San Joaquin Delta College, Stockton

Paramedic Training; EMT-1A Training – City College of San Francisco
Basic Life Support & Advanced Cardiac Life Support Certificates – American Heart Assn.
Ambulance Driver's License – Advanced Airway Management Training
Additional courses in Anatomy, Physiology, Chemistry

ROBERT LAWTON

12345 Wandering Road
Happy Valley, CA 94546
(707) 321-4567

OBJECTIVE

Part-time position as service writer for an auto manufacturer.

SUMMARY

- Eight years' experience in automotive diagnosis and repair.
- Able to assess customer's needs and earn their confidence and trust.
- Good judgment in balancing customer service with profitability.
- Work cooperatively with a wide range of personalities.

RELEVANT EXPERIENCE

Customer Service

- Generated a large volume of repeat business from satisfied customers, maintaining excellent relations through good repair work and sensitivity to clients' overall needs.
- Developed a sharp awareness of both the spoken and unspoken needs of customers, providing adequate technical information and advice for decision making, and restoring customers' sense of "being in control."

Technical Knowledge

- Completed 2 years training in automotive repair at Chabot College.
- Repaired foreign and domestic cars for 8 years, specializing in:
  –electronic fuel injection –SU carburetors –engine rebuilding.
- 5 years' experience in heavy machinery operation and repair.

Business Skills

- Started up a successful restaurant business, including leasing / remodeling the building; hiring and supervising staff.
- Created freelance auto repair business, handling all aspects of operations

EMPLOYMENT HISTORY

1991–present	**Freelance Part-time**	ROBERT LAWTON FIXES CARS, Davis, CA
1990	**Mechanic**	WINTERS GARAGE, Winters, CA
1988–89	**Mechanic**	J & J AUTOMOTIVE, Davis
1982–87	**Machine Operator**	BOSTROM BERGEN METAL PRODUCTS, Oakland
1980–82	**Partner**	ALL ONE NATURAL FOODS, Hayward

EDUCATION & TRAINING

Economics major – U.C. DAVIS and HAYWARD STATE, 1990–present (part-time)
Certificate in Auto Mechanics – CHABOT COLLEGE, 1988

Additional Training:
Electrical Systems; Tune-up – GENERAL MOTORS TRAINING CENTER
Electronic Ignition; Infrared Diagnosis – SUN PRODUCT TRAINING, San Jose

Compare this version of Stephen's resume with the version on page 48. His experience is reorganized, with skill headings re-named to fit each different job target.

STEPHEN R. HONDA

788 Manada Avenue • Oakland, CA 94612 • (510) 890-6443

Objective: Position as real estate analyst/researcher for a major development firm.

SUMMARY

- Conducted extensive research and analysis of RE prices and availability.
- Designed and implemented surveys on relocation needs.
- Adept in counseling residents on relocation options and responsibilities.
- Strength in innovative program planning, funding, and analysis.

REAL ESTATE EXPERIENCE

Research and Analysis

- Documented vacancy rate in Oakland neighborhood: made door-to-door survey in a high-density neighborhood to determine the vacancy rate in the area.
- Researched local condominium prices for the city of Concord:
  - Visited and viewed new condominiums in the community;
  - Gathered accurate information on current market prices and made comparison to those proposed by developer seeking municipal assistance.
- Designed and implemented a state-wide survey of city-funded rental rehabilitation programs:
  - Searched files to identify cities with rental rehabilitation programs;
  - Designed questionnaire to identify and describe those programs with displacement protection;
  - Wrote comprehensive report summarizing the programs that featured protection.
- Researched legal procedure for transferring property management responsibility of abandoned rental properties; published results in Renters' Resource Manual.

Writing & Presentation

- Mapped results of apartment vacancy survey in Oakland, and made a formal presentation to City Council.
- Wrote detailed report advising City Attorney of New York City on procedure for complex land transfer and write-down for Section 8 rental development.
- Authored concise information bulletin on city of Concord Mortgage Assistance Program.

EMPLOYMENT HISTORY

1994–present	*Classroom teacher*	Oakland and Richmond public schools
1993	*Office Manager*	Pacific Car Rental, Oakland
1991 (summer)	*Administrative Asst.*	**Community Development Dept.**, Concord
1990–1992	Full-time grad student	Columbia University, NYC
1990	*Researcher/Writer*	**Task Force on City-owned Property,** NYC
"	*Library Clerk*	UC Berkeley
1989	*Housing Advocate*	**Oakland Citizens Committee for Urban Renewal**
1988 (summer)	*Housing Intern*	**Calif. Dept. of Housing & Community Development**

EDUCATION & TRAINING

Completed graduate course work in Urban Planning, COLUMBIA UNIVERSITY, NYC
B.A., Geography – UC BERKELEY

DONALD RAULINGS

Private Investigator

License No. AQ 987654

1219 Herrick Court ■ Oakland, CA 94609 ■ (510) 658-9999

SUMMARY OF QUALIFICATIONS

- 17 years' professional experience in law enforcement; extensive contacts in the law enforcement field.
- Extensive specialized training in all aspects of investigation, including fire cause, fraud and surveillance.
- Experience in police internal affairs and preparation of civil liability cases, requiring objectivity and integrity.
- Persistent, thorough and prompt in completing projects.
- Excellent communication and interrogation skills; reputation as one of the best interviewers on staff.

RELEVANT PROFESSIONAL EXPERIENCE

INVESTIGATION

Conducted hundreds of complex investigations of major crimes, in 6 years as Oakland Police Department Detective:

- Prepared detailed investigative reports, resulting in criminal prosecution and conviction.
- Successfully elicited confessions for such crimes as homicide, robbery and arson, applying communication and interrogation skills.
- Investigated complaints against police officers, internally and externally generated, during 2-year assignment as Internal Affairs Sergeant reporting directly to Chief of Police.
- Prepared civil liability defense claims against individual police officers and City of Oakland.

CASE MANAGEMENT

- Managed investigative projects independently, applying sharp analytic and problem solving skills, exhausting all available leads.
- Planned, organized and supervised inter-jurisdictional recovery system dealing with crimes of burglary and possession of stolen property.
- Developed and coordinated a task force of police officers from several communities.

SECURITY CONSULTATION

- Developed original Crime Prevention Program for city of Oakland. Conducted security evaluation surveys, both residential and commercial, providing solutions to security problems.
- Provided retail security management for Sears, Oakland, auditing cash flow and credit receipts, and monitoring employees, in order to maintain internal and external theft-loss control.

– Continued –

DONALD RAULINGS
Page Two

Don's professional independence is clear at the top of the resume. Note that even though his resume covers two pages, the entire skill assessment is complete on page one.

EMPLOYMENT HISTORY

1996–present	**Private investigator**	RAULINGS INVESTIGATIVE SERVICES, Oakland
1994–96	**Detective Sergeant**	OAKLAND POLICE DEPARTMENT
	Acting Superintendent	" " xAnimal ControlBureau
1991–94	**Sergeant**	" " Internal Affairs, Patrol Division
1986–90	**Detective**	" "
1979–86	**Police Officer**	" "
1982–84	**Security Agent,** part-time	SEARS ROEBUCK CO, Oakland
1978–89	**Machine Operator**	MILLER IRON WORKS, Oakland
1976–78	**Petty Officer, 3rd Class**	U.S. NAVY; honorable discharge

INVESTIGATIVE & MANAGEMENT TRAINING

- Completed advanced investigative training in a wide range of specific crimes:
  - Robbery Investigation; San Jose State
  - Burglary Prevention Seminar; Cal State/Long Beach
  - Auto Theft School: Advanced Auto Theft; Dept. of Justice, Sacramento
  - Homicide Investigation; Justice Training Center, Sacramento
  - Homicide Investigation; Oakland Police Dept.
  - Arson for Profit; State Fire Marshall, Columbia Jr. College
  - Advanced Arson Investigation; State Fire Marshall, Columbia Jr. College
  - Internal Affairs Investigation; Cal State/Long Beach
  - Recognition & Apprehension of Narcotics and Drug Offenders; Oakland Police Dept.

- Completed courses in technical and management skills:
  - Evidence Technician and Photography; Bahn Fair Institute of Scientific Law Enforcement
  - Investigation School; Los Angeles Police Dept.
  - Stress Management; Cal State/Long Beach
  - Family Violence and Child Abuse; Chabot College
  - Performance Appraisal; Oakland Police Dept.
  - Defensive Tactics; Oakland Police Dept.

EDUCATION & CREDENTIALS

B.S., Administration of Justice; minor in Sociology – SAN JOSE STATE
A.A., Administration of Justice – CHABOT COLLEGE, Hayward

California Police Officer Standards Training: Basic, Intermediate, Advanced, Supervisory Certification

References available on request.

TERESA FERNANDEZ

3556 - 8th Avenue ◆ San Francisco, CA 94118
(415) 808-6578 email: terfer@global.com

Objective: Position as wardrobe assistant or supervisor with a movie company or TV soap opera.

SUMMARY OF QUALIFICATIONS

- Able to handle a multitude of details at once, under pressure and deadlines.
- Effective in managing dressing room environment with diplomacy and authority.
- Well organized, punctual, resourceful; can be counted on to get the job done.
- Successful experience in wardrobing for international rock tour.
- Work equally well as a team member or independently.

WORK EXPERIENCE

1995–present **Wardrobe Mistress** — ROADHOUSE, rock band, New York City

MANAGING DRESSING ROOM

- Monitored dressing room environment for a rock group of seven traveling throughout the US, Canada and Europe to assure safety, efficiency and privacy:
  – Hosted backstage guests and business associates of the band;
  – Assured proper security during shows;
  – Cleared dressing rooms of unauthorized people before/after/during shows.
- Coordinated with caterers on delivery and set-up of contracted food and beverages, assuring that contract agreements were kept; advised caterers on special dietary needs.
- Supervised physical exercise class for the band.

APPOINTMENTS/LOGISTICS

- Scouted directions at each new gig, and advised band members on location of showers, rest rooms, dressing rooms, dining facilities, stage locations, production office, promoters' and reps' names and office locations.
- Located hairdressers and masseurs on the road and arranged on-site appointments.

COSTUME MAINTENANCE

- Selected and laid out costumes for concert appearances.
- Maintained costumes:
  – Pressed stage costumes with electric iron and portable steamer;
  – Spot cleaned clothing; polished shoes; cleaned and repaired jewelry.
- Packed wardrobe cases for daily transit, handling clothing, accessories and makeup.

BOOKKEEPING, SHOPPING, ERRANDS

- Located and bought stage accessories: shoes, belts, gloves, stockings, hats, jewelry.
- Bought and replaced makeup and toiletries; kept detailed records of purchases.
- Located facilities in different cities each day, for laundry, dry cleaning and shoe repair.

– Continued –

See Teresa's cover letter on page 267.

TERESA FERNANDEZ
Page two

EMPLOYMENT HISTORY

July 1996	**Wardrobe**	HBO film, "The Hustler's Daughter"
1994–96	**Freelance Designer Asst.**	EUGENE STEPHENS DESIGNS leather accessories – NYC
1994–95	**Freelance Designer Asst.**	SUZANNE WORTHINGTON quality women's wear – NYC
1993–94	**Designer's Rep**	OLIVIERI OLIVIERI women's clothing/accessories – NYC
1991–92	**Retail Sales/Bookkeeping**	a women's boutique; a film/video supply store
1988–92	Student and apprentice in acting and writing comedy for stage and video; concurrent with part-time jobs	

EDUCATION & CREDENTIALS

B.A., Art, with honors – SAN FRANCISCO STATE UNIVERSITY

POTPOURRI

KENNETH H. RICHARDS

1977 Hastings Way
Hayward, CA 94703
(510) 949-6734

OBJECTIVE

Warehouseman.

SUMMARY

- Over 15 years' experience in warehousing; 10 years supervision.
- Highly effective and positive approach to supervising others.
- Head of Foghorn Returns Department for 10 years. Entrusted to handle this unit without direct supervision.
- Retained customer goodwill through timely processing of returns saving the company hundreds of thousands of dollars annually.

WAREHOUSE EXPERIENCE

1981–1996 **FOGHORN PRESS Berkeley, CA**
Returns Supervisor/Warehouseman

- Supervised 1-5 other workers regularly. Periodically oversaw up to 15 other workers of various cultures and languages.

- Taught workers, through my own experience, to:
  - deal with the daily challenges of the job in a constructive way;
  - recognize their own accomplishments; - work efficiently;
  - maintain morale and motivation.

- Developed an effective strategy for minimizing worker turnover, despite seemingly impossible conditions.

- Single-handedly coped with high volume of returns.

- Successfully set up a system to absorb a 200% increase in volume with no increase in manpower for the first 8 years.

- Experience in all areas of warehouse work:
  - Forklift operation - Shipping and receiving
  - Loading/unloading - Freight company interaction
  - Invoice fulfillment

1978–81 **CHEMCO PHOTO PRODUCTS**
Warehouseman (1 year)
Warehouse Supervisor (1-1/2 years)

1974–1978 **EDUCATIONAL TESTING SERVICE,** Berkeley, CA
Mailroom

Carlene's "Profile" is very direct, confidant, and personalized, reflecting her sense of her own value to an employer.

CARLENE DOONAN

2731 Southgate Drive • Berkeley, CA 94702 • (510) 525-9934

OBJECTIVE **A position as apprentice baker,** with opportunity for quality training and increasing levels of responsibility.

PROFILE

- Work hard, learn fast, willing and able to assume responsibility.
- Studied cooking with a Master Chef.
- Passion for food; commitment to producing highest quality products.
- Experience with successful retail design and display.
- Good team player; work well with all kinds of people.

EXPERIENCE

Cooking Knowledge

- Studied with master chef, Ken Wolfe, learning:
  - Principles and techniques of food preparation;
  - Importance of quality and freshness of ingredients;
  - Chemistry and effects of combining ingredients;
  - Coordinated timing of food preparation;
  - Innovative approaches to traditional cooking principles;
  - Balancing flavors within a dish and within a meal.

Speed/Accuracy

- Prepared puff-pastry dough in cooking class, consistently completing in record time.
- In other business settings:
  - Handled large volume of theatre customers in minimum time;
  - Accurately counted, recorded and deposited retail cash receipts.

Coordination/Teamwork

- Maintained and supervised a balanced flow of inventory for theatre food concession and for retail gift store:
  - Monitored sales including seasonal fluctuations;
  - Researched to determine best prices, by phone and at trade shows;
  - Assured correct stocking and display.
- Coordinated timing and priority of tasks, as store manager.
- Worked on a finely tuned sales team to expedite customer services.

WORK HISTORY

1988–present	**Store Manager**	MIASMA gift store, Berkeley
1985–88	**Assistant. Mgr.**	" " "
1984–85	**Theatre Manager**	RENAISSANCE-RIALTO THEATRE GROUP Berkeley
1983–84	**Clerk/Counter Sales**	" " "

EDUCATION Laney College, 1982–85, Liberal Arts

Classes in Cooking Principles with Ken Wolfe, Master Chef

POTPOURRI

DAWN ELLSWORTH

1415 Montgomery St.
Oakland, CA 94611
(510) 548-2223

Objective: Gallery assistant or gallery sales position.

SUMMARY

- Demonstrated ability to communicate knowledgeably with clients about specific artists, styles, techniques and media.
- Able to represent a studio or gallery in public presentations.
- Effective in sales through enthusiasm for the product.
- Planned, managed and supervised events for up to 1,000 people.

RELEVANT EXPERIENCE

CONSERVATION & MUSEUM SERVICES

- Conserved works of art and documents at private conservation studio, involving: ...washing ...cleaning ...flattening ...mounting ...matting ...framing ...aesthetic restoration.
- Evaluated damage to art pieces and documents and proposed individual treatments.
- Selected appropriate museum-quality ph-balanced materials to frame and mat both old and new pieces.
- Researched and compiled the provenance of recently acquired paintings at SFMMA.
- Inventoried and registered museum's permanent collection for future catalogue.

ART MEDIA & DESIGN

- Developed working experience with wide range of media and techniques: ...glass ...painting ...textile design ...pastel and graphite.
- Designed and produced new stained glass pieces, and restored old stained glass.

PRESENTATION / SALES

- Assembled and presented slide show on art history and restoration techniques.
- Advised art clients of inventory and pricing; made appointments for sale closings.
- Answered technical and historical inquiries from potential clients by phone and letter.

COORDINATION

- Coordinated and supervised 5–50 staff members for large catered events: delegated staff assignments; mediated between party hosts, guests and staff; handled complex logistics for all equipment.

WORK HISTORY

1995–present	***Party Manager***	MARSHALL CATERING, Oakland
1993–1995	***Conservation Asst.***	KAREN ZUKOR (conservator of art on paper), Oakland
1993–1994	***Painter***	FRALEY STUDIOS (carousel restoration), Oakland
1992–1993	***Waitress***	GRIFFON RESTAURANT, Berkeley; LITTLE JOE'S, S.F.
1990–1991	***Full-time student***	U.C. Santa Cruz
1988–1989	***Glass Designer***	Self-employed, Los Angeles

EDUCATION & TRAINING

Museum Management & Exhibition Design, JFKU Center for Museum Studies (current)
Italian restoration techniques, Inst. of Art & Restoration; Florence, Italy, 1994
B.A., Fine Arts (Painting & Print Making) – U.C. Santa Cruz, 1991

Hellmut lists his earliest jobs first to focus attention on his experience in warehousing, even though this experience was gained many years ago. See Hellmut's other resume on pages 16–17.

HELLMUT DIETRICH

333 - 65th Street, Oakland, CA 94609
(510) 699-4742

Objective: Position in import-export business using background in freight handling.

PROFILE

- Experience in office work with a major freight company.
- Competent in operating computers and programming.
- Skilled supervisor; able to motivate others and handle conflict.
- Sharp and creative in solving problems.
- Diplomatic and effective with customer relations.

EXPERIENCE

Warehouse • Freight • Dispatching

- Supervised the loading and unloading of goods, assuring that items were handled with care and placed accurately in warehouse.
- Developed and updated time schedules and delivery tours for truckers.
- Initiated improvements in teamwork efficiency in freight warehouse, by clarifying areas of responsibility and promoting communication between office and warehouse.

Computer Skills

- Familiar with use of personal computers and printers.
- Customized commercial computer programs to meet special needs.

Customer Service

- Handled inquiries and complaints of 3 major manufacturing customers, as warehouse superintendent at a large freight company in Munich.
  – Took orders for freight service and provided complex pricing information;
  – Resolved problems involving shipping delays and damage.

EMPLOYMENT HISTORY

1977–80	***Warehouseman***	HOHL FREIGHT CO., Michelfeld, Germany
1980–81	***Warehouse Supv.***	ANTON GLATZ FREIGHT CO., Munich, Germany
1982–86	***Student***	City of Schwäbisch Hall, Germany
1983–85	***Teaching Asst.***	PÄDAGOGISCHE HOCHSCHULE COLLEGE, Esslingen, Germany
1986–88	***Teacher***	Jr. High School, Elementary School; Osfildern, Germany
1989–93	***Display Builder***	S&E STARK PROMOTIONAL DISPLAY CO., Ostfildern, Germany
1994–95	***Student***	UNIVERSITY OF TÜBINGEN, Germany
1996–present		Travel and relocation to the Bay Area

EDUCATION

B.A., Social Work (German equivalent) – University of Tübingen, Germany
Classes in: Bookkeeping, correspondence, transportation, business and math.

DOLORES L. WALKER

59 Orchid Drive • Mill Valley, CA 94947 • (415) 644-9889

OBJECTIVE: Position as commercial leasing agent.

SUMMARY

- CEO of a $17 million business; ten years' executive experience.
- Outstanding ability to build trust and deal effectively with people.
- Adept at tuning in to clients' priorities, accurately assessing needs.
- Proven successful in negotiating contracts and leases.

PROFESSIONAL EXPERIENCE

— As Executive Director of ASUC —

Sales & Marketing

- Increased annual sales from $11 million to $17 million in 3 years.
- Initiated new marketing unit to improve organizational image and raise development funds.
- Persuaded board of directors to adopt a 5-year plan, after history of crisis-oriented management.
- Assessed needs of departments, units and individuals to improve working environment.

Business Contacts

- Developed extensive potential client base through business contacts with attorneys, bankers, professional organizations, entertainers' agents, merchants, alumni.

Contract Negotiation

- Successfully renegotiated long-standing unfavorable contract with major vending contractor:
  – Increased commission schedule 40%; – Removed exclusivity clause;
  – Significantly improved level of services to campus community.
- Negotiated contracts with a wide diversity of businesses: ...food service ...bank ...entertainment ...hair salon ...travel center ...stationery store.
- Maintained rapport and professional effectiveness with vendors under conditions of political diversity and ever changing demands.

— As Office Manager, UCSF Material Management Dept. —

Facilities Management

- Coordinated provisions for facilities of major service and commercial complex: ...building maintenance ...custodial service ...energy management ...HVAC systems.

EMPLOYMENT HISTORY

1992–present	**Executive Director**	Associated Students (ASUC), Univ. of Calif.
1985–92	**Asst. Exec. Director**	" " "
1978–85	**Office Manager**	U.C. San Francisco, Material Management Dept.
1975–77	**Sales**	Self-employed specialty clothing distributor

EDUCATION

Business and Legal studies – Armstrong University, Berkeley
Classes in Psychology, Financial Accounting, Fund-raising
Completed licensing course – Anthony School of Real Estate
Xerox Professional Selling Skills System III

See Denise's cover letter on page 268.

DENISE FRANCIS

6200 Skyline Blvd. • Oakland, CA 94611 • (510) 754-5151

OBJECTIVE

Position as a cruise staff member working directly with passengers, such as activity director, hostess, gift salesperson, tour escort, deck steward, masseuse.

HIGHLIGHTS

- Successful work experience in activity directing, child care, nursing, massage, and sales.
- Enthusiastic, fun loving and personable.
- Outstanding leadership and organization skills.
- Honest and reliable in keeping commitments.
- Relate easily and openly with all ages and types.

SKILLS AND EXPERIENCE

Activities Directing

- Planned and supervised special event parties for 100 people, including weddings, bon voyage parties, birthdays:
  – Designed floral and food arrangements
  – Installed table and wall decorations – Planned menu
  – Coordinated schedule – Supervised outdoor play activities.
- Successfully involved convalescent adults in healthful activities, persuading them to participate in coffee hours, arts-and-crafts activities, and exercise.
- Supported the health and happiness of home-bound patients through assistance with entertainment, personal care and recreation.
- Taught a highly popular class in massage therapy.

Promotion / Sales

- Developed friendly, supportive, give-and-take relationships with coffee shop patrons, building a loyal base of repeat customers.
- Promoted and sold merchandise in a variety of settings:
  – flower shop – coffee shop – gift shop

Interpersonal Communication / Needs Assessment

- Completed classes in psychology, assertiveness and interpersonal communication.
- Assessed and worked with physical and emotional stresses of clients, as massage therapist and home health attendant.

WORK HISTORY

1993–present	***Home Health Attendant***	Independent Contractor, Oakland
1993–94	***Floral Designer***	SIEFERT'S FLORAL, Oakland
1990–96	***Masseuse & Teacher***	Independent Contractor, Oakland
1991–93	***Assistant Manager***	THE IRIS, arts gift shop, Berkeley
1989–91	***Assistant Manager***	THE COFFEE MILL, Oakland
1984–88	***Nurses' Aide***	CLARK ADULT HOME, Oakland

EDUCATION

Merritt College, Oakland, 1991-96, on-going degree studies
Certified Activities Director; Nurse Assistant; Floral Designer
Certified Masseuse – Massage Institute, San Francisco
Certified Independent Travel Agent – Tivoli Travel School, SF

Youth and Student Resumes

Matthew Kurle (age 8)	Nothing listed yet	257
Julia Smith (age 14)	Baby-sitting/odd jobs	258
Charlie Knych	After-school jobs (high-school student)	259
Karen Schmidt	Marketing trainee (college student)	260
Valerie Lauer	Entry, office services (college-bound)	261
Rachael Hennesey	Internship, public law (college student)	262

At 8 years old, Matthew gets his first taste of what the grown-ups call "skill assessment."

My First Resume

Matthew Kurle

19055 James Blvd.
Poulsbo, Washington 98370
(360) 123-4567

Job I want when I grow up:
Painting at my Dad's Shop

My skills

Basketball
Baseball
Soccer
Tetherball
Making change
Writing
Swimming
Running
Math
Reading
Art
Telling time
Badminton
Using two computers
Making things to sell
Selling things
Painting pictures and coloring
Making things with clay
Using Grandma's copy machine
Taking pictures with my camera

Ways I Know How to Make Money

Doing my chores
Making paper things and selling them
Selling chocolate candies for my school
Having a bank account

Things I Want to Learn in School and College

How to work
How to drive a car
How to be a doctor
How to drive a motorcycle
How to make books
How to fix cars
9th grade math
How to do plumbing

Julia's only 14, but she already knows several ways to make money.

JULIA SMITH

Please contact me at (604) 987-6543, during the hours of 8am–7pm.

SERVICES I CAN OFFER:

- Baby sitting
- Lawn mowing
- Raking leaves
- Car washing
- Pet cleaning & grooming
- Pet sitting (feeding & walking)

Baby Sitting:
For the last seven years I have taken care of children from age 10 months to 10 years old. Responsibilities included: cooking; clean-up; feeding and clothing children; ensuring children are home at the correct time; providing breakfast and making bagged lunches; and getting children ready for school.

Car Washing:
I am careful when washing cars and make sure that both the inside and outside of the vehicle are clean, vacuumed and dusted.

Lawn Mowing:
Last summer I earned all of my spending money by mowing lawns. I do a thorough job of mowing, raking and weeding lawns.

Pet Care:
I am experienced in bathing dogs and cats. I love animals and have a dog and three cats, and have washed and clipped my golden lab-retriever. I enjoy pet "sitting" both dogs and cats.

Raking Leaves:
I have some experience raking leaves, and I love autumn.

VOLUNTEER EXPERIENCE:

1994–95: **Volunteer Teacher's Assistant**, Ms. Jean Richards' Grade 5 Class, Smithton Elementary School, Campbell River, B.C.
Assisted students with reading, writing and math lessons.

1991: **Volunteer Model,** Modern Hair Salon, Campbell River, B.C.
Modeled hair styles.

INTERESTS:

Water and snow skiing, other outdoor sports. Participated in Grade 3-8 Track and Field activities. I enjoy gymnastics and choir, also.

– Resume written by Cory Beneker –

Charlie felt more confidant about job hunting once he got his skills down on paper.

CHARLIE KNYCH

5112 N. Avenida Primera • Tucson, Arizona 85704 • (602) 887-6190

Job wanted: Part-time after school and weekend job
...Stockroom helper ...Grocery bagger ...Golf Caddie ...Dishwasher
...Theater maintenance helper ...Arcade guide

- Reliable. Willing and able to show up on time.
- Honest and trustworthy.
- Good attitude around others. Willing to help, patient with people, do my share, willing to learn.

Paid Work Experience

1993–94 **Handyman**
- Worked on Tucson rental property owned by my relatives.
  - ... Replaced broken wood fence.
  - ... Painted inside and outside walls.
  - ... Cleared out trash left by tenants.

Fall 1994 **Babysitter**
- Do on-call work for parents during football games.

1992–93 **Bagger** after school, weekends, vacations
- Bagged groceries at US Army Commissary, Wurzburg, Germany

1991 **Woodworker and Salesman** part-time during school year
- Produced wood handicrafts; sold them to teachers and other students.
  - ... Increased sales by giving customers a chance to try out the merchandise and show it to others.
  - ... Increased income by raising prices on popular items.
  - ... Used woodworking tools to make handicrafts.

1992 **Golf Caddie**
- Caddied every day during summer vacation.

Work Experience in School

Animal Care Project, Nov. 1992 to Apr. 1993
- Successfully raised a pig to sell at the County Fair.
  - ... Won three blue ribbons. Earned over $200.
  - ... Fed and washed the pig every day, cleaned the pen, gave the animal shots when it was sick, checked its weight weekly.

Free Enterprise Class, 1990 (Junior High)
- Sold school supplies such as pencils, pens, paper, notebooks.
- Designed, built, and sold wooden plaques which were popular with teachers.

EDUCATION: will graduate from high school in June 1995

KAREN D. SCHMIDT

School Address:
11305 N. University, Apt. B
Peoria, Illinois 61606
(309) 987-6543

Home Address:
612 Dacker Ridge
St. Louis, Missouri 63129
(314) 876-5432

OBJECTIVE

Marketing Trainee

SUMMARY

- Excel at balancing leadership activities and academic responsibilities.
- Developed marketing skills in public relations firm.
- Strengthened communication abilities through leadership activities.
- Extensive exposure to customer relations.

EDUCATION

Bachelor of Science in Marketing
Bradley University, Peoria, Illinois
Expected date of graduation: May 1997
Major GPA: 4.0 Overall GPA: 3.7

AWARDS & HONORS

Mortar Board, Senior Honorary Society, 1995–present
Beta Gamma Sigma, National Junior Honor Fraternity, 1995–present
Phi Eta Sigma, National Freshman Honor Fraternity, 1994–present
Dean's List (4 semesters)
E. A. McCord Scholarship, 1994–1996

PROFESSIONAL EXPERIENCE

Summer Intern, June 1995–August 1995
Clayton-Davis & Associates, Inc., St. Louis, Missouri
- Developed public relations skills through media coordination, copy editing, and research.
- Composed articles for national publication.

Administrative Intern, July 1994–August 1994, May 1995–August 1995
D & C Forms, Inc., St. Louis, Missouri
- Created graphics for customized forms.
- Managed receipt of purchase orders.
- Devised a simple systematic filing system.

CAMPUS EXPERIENCE

Student Activities Budget Review Committee, 1995–present
Student Advisory Committee, Business College, 1995–present
American Marketing Association, 1994–present
Phi Chi Theta, Business Fraternity, 1994–present
Activities Council of Bradley University, 1994–95
- Co-Coordinator of Homecoming and Fall Fest Events

– Resume written by Karen Schmidt –

VALERIE LAUER

6808 Chambers Drive
Oakland, CA 94611
(510) 339-3022

Job Objective: Entry position in office services.

SUMMARY OF QUALIFICATIONS

- Eager, hardworking, and reliable.
- Willing to learn and accept constructive criticism.
- Highly motivated for career advancement.
- Enjoy contributing to a team effort.
- Help create a pleasant, clean working environment.

OFFICE EXPERIENCE

GENERAL OFFICE SKILLS

- As office assistant:

– Answered phones	– Filed documents	– Proofread
– Made appointments	– Typed correspondence	– Ran errands

BUSINESS SKILLS

- Entered orders into FileMaker database.
- Made and tracked invoices, ensuring satisfaction and payment.
- Calculated and handled payments by check and cash.
- Filled customer orders for books and software.

COMPUTER KNOWLEDGE

- Basic knowledge of Macintosh software (FileMaker, Word).
- Familiar with IBM software (Word).

EMPLOYMENT HISTORY

1994–present	**Office Assistant**	Damn Good Resume Service, Oakland, CA
1994–present	**Baby-sitter**	Witherall House, Oakland, CA
Summer 1994*	**Animal Care**	Children's Fairy Land, Oakland, CA

(*volunteer job)

EDUCATION

Skyline High School, Oakland, CA, graduated June 1996

– Resume written by Valerie Lauer –

RACHAEL E. HENNESEY

9876 Carmel Way
Fairfield CA 94533
(415) 123-4567

Objective

Internship position with a private or public law agency, providing experience and exposure in the field of criminal law.

Highlights

- Strong commitment to a universally available and equitable legal system, working within the criminal law system to achieve this end.
- Completed extensive college coursework in law.
- Intelligent and self-motivated; maintained a 3.8 GPA while holding down several jobs and volunteering in community service.

Relevant Experience

1993–present — ***Full time student, Mills College***
Relevant coursework:
- Law & Society •Legal Aspects of Business •Psychology & Law
- Social Inequality •Court Systems of the Metropolitan Area
- Public Policy Making Process

1994–present — **BAWAR** (Bay Area Women Against Rape)
Crisis Counselor – 10 hours per week
- Provided hot-line counseling 4-5 times monthly in 4-hour shifts:
  – Developed acute listening skills;
  – Probed and identified problems;
  – Provided information on community resources;
  – Advised on legal procedures and rights;
  – Interacted with law enforcement and medical agencies.

Aug '94–May '95 — **MILLS COLLEGE**
Aug '95–May '96 — ***Residential Assistant***
- Counseled students on academic, personal, and crisis issues.
- Assessed students' interests and developed social and academic programs; arranged for speakers and logistics for these programs.
- Managed and regulated student residence hall.

1994–present — **PACIFIC GAS & ELECTRIC**
Engineer's Aide / Field Clerk
(Full-time summers '94 and '95; part-time current academic year)
- Operated both Macintosh and IBM computers, implementing a complex computer test generation program.
- Familiar with a variety of software, including:
  –WordPerfect –Microsoft Word –VolksWriter –SuperCalc
  –Dreams –Newsmaster –Labels –Ability.
- Researched apprenticeship training files.
- Scheduled instructors' classes.
- Co-developed an organizational structure for the entire office.

Education

B.A. in progress (1997) – Mills College, Oakland
Major: **Political, Legal, & Economic Analysis** • Minor: Women's Studies

COVER LETTERS

Sample Cover Letters

Letter Page	Name	Resume Page
264	Tricia	189
265	Richard	154
266	Joyce	204
267	Teresa	248
268	Denise	255
269	Marsha	35
270	John	104
271	Donna	184

Your resume should always have a good COVER LETTER attached, as a personal communication between you and the individual who receives the resume.

Most people are intimidated by this task, but it's not that hard if you think of it as just a **friendly, simple communication from one person** (who's looking for a good job) **to another** (who's looking for a good employee). It is in the interests of both parties to make a good connection!

How To Write a Good Cover Letter

1. **Be sure to address it**—by name and title—**to the person who could hire you**. When it's *impossible* to learn their name, use their functional title, such as "Dear Manager." You may have to guess ("Dear Selection Committee") but *never* say "To whom it may concern" or "Dear Sir or Madam"!
2. **Show that you know a little about the company**, that you are aware of their current problems, interests, or priorities.
3. **Express your enthusiasm and interest** in this line of work and this company. If you have a good idea that might help the employer resolve a problem currently facing their industry, offer to come in and discuss it.
4. **Project warmth and friendliness,** while still being professional. Avoid any generic phrases such as "Enclosed please find..." *This is a letter to a real live person!*
5. **Set yourself apart from the crowd.** Identify at least one thing about you that's unique—say a special talent for getting along with everybody at work, or some unusual skill that goes beyond the essential requirements of the position—something that distinguishes you AND is relevant to the job. (Then, if several others are equally qualified for the job, your uniqueness may be the reason to choose YOU.)
6. **Be specific** about what you are asking for and what you are offering. Make it clear which position you're applying for and just what experience or skill you have that relates to that position.
7. **Take the initiative** about the next step whenever possible, and be specific. "I'll call your office early next week to see if we could meet soon and discuss this job opening," for example. OR—if you're exploring for UN-announced jobs that may come up—"I'll call your office next week to see if we could meet soon, to discuss your company's needs for help in the near future."
8. **Keep it brief**—a few short paragraphs, all on one page.

See Tricia's resume on page 189.

TRICIA BAKER

9490 Wheelwright Road • Clio, MI 48409 • (810) 876-5432

David L. Becker
CLU 4800 Fashion Sq. Blvd.
Suite 500
Saginaw, MI 48604

Dear Mr. Becker,

My name is Tricia Baker and I am on the lookout for a new position. After ten years of business management, I am aware of the difficulties many employers have in choosing the right person for their company. We ask ourselves if this person will learn the job quickly, will they be loyal and reliable, and most importantly, will they help our company save time and money? Did you know that studies have shown that your chances of choosing the right individual for the job are only 3% better than if you had chosen that person's name out of a hat?

When you need a new person to fill a position on your team and would like to take the guesswork out of your hiring, please consider my qualifications. With three years of experience in the Graphic Arts and Screen Printing Industry from 1992–1995 and ten years of experience in Retail Management and Sales from 1982–1992, I have a well rounded background. I am a person who is resourceful and creative, self-motivated, takes lots of initiative, enjoys a challenge and has great physical stamina.

I will be relocating to the area in mid-December and would like an opportunity to meet with you in person to discuss your personnel needs and present some of my experience and accomplishments. Next week I will follow up with a call to your office in hopes that this can be arranged.

Sincerely,

Tricia Baker

– Cover letter written by the job hunter –

See corresponding resume on page 154-155.

RICHARD JENNINGS
8009 Mountain Blvd.
Oakland, CA 94602
(510) 234-5678

James Lawler
Director of Operations, Alameda County
Bay Medical Services
780 San Joaquin Street
Oakland, CA 94604

Dear Mr. Lawler,

I was very pleased to learn of the opening for the Alameda County Field Supervisor-Operations position.

In my 18 months as a paramedic with Bay Medical Services, I have demonstrated independence and initiative in developing plans to resolve the problems I encountered on the job.

As a Crew Chief and Relief Field Supervisor, I have enjoyed participating in the operational functions of BMS.

I look forward to opportunities to take on even more responsibility in this organization — and I see the Supervisor-Operations position as a chance to further contribute to the development of paramedicine, as well as to Bay Medical Services as an organization.

A copy of my resume is enclosed. I plan to call your office next Monday to inquire about when we could meet, and I look forward to speaking with you in person about this position.

Sincerely,

Richard Jennings

See corresponding resume on page 204.

JOYCE STROEBECH
578 Willow Drive
Walnut Creek, CA 94598
(510) 902-1228

Marilyn Sneider
Advertising Director, KARAN
9000 Maiden Lane
San Francisco, CA 94107

Dear Ms. Sneider:

I recently visited the KARAN offices in San Francisco and was immediately impressed by your positive and aesthetic environment for creativity. I especially connected with the inspirational messages mounted on the walls. Since then I have read many articles about KARAN and have talked to several people about the company, including Jean Livingston in the International Dept. I am very impressed with your current ad campaign and its strong impact of color, simplicity, personality, and energy.

I knew from the beginning that I wanted to support the KARAN movement, to be a dedicated and integral part of the company. I believe very strongly in the KARAN commitment to individual style, fitness, fun, and originality.

I have many skills and experiences gained from my years of work in design, retail and wholesale, and management, as outlined in my resume. I have a good color sense, an eye for coordinating fabrics, fashions, and accessories, and a playful attitude. I'm a resourceful investigator, creative problem solver, and a strong motivator. I like to research new trends, analyze information, and discover new inspirations. I believe my experience and enthusiasm will make me a valuable team member helping KARAN continue to grow and inspire.

I would like to talk to you in person and discuss where my skills would benefit you the most. I am looking forward to hearing from you soon. The best time to reach me is between 9 AM and noon, at (510) 902-1228.

Sincerely,

Joyce Stroebech

See corresponding resume on page 248-249.

TERESA FERNANDEZ
3556 - 8th Avenue
San Francisco, CA 94118
(415) 808-6578 email: terfer@global.com

May 12, 1996

Kathy Nishimoto
Manager, Wardrobe Dept.
CBS-TV
7800 Beverly Blvd.
Los Angeles, CA 90036

Dear Ms. Nishimoto,

In late October I will be in Los Angeles checking out job possibilities in my field of wardrobe, since I will be moving to the area soon.

I would particularly like the chance to meet with you to discuss the possible use of my skills and experience at CBS. I'm especially enthusiastic about working with soaps such as "Marlborough Heights" or "Secret Passions," the show I love and never miss.

You will see on my enclosed resume that I've successfully handled full responsibility for the wardrobe of a band of 7 people doing live performances around the world. I've also had experience with wardrobe for film and video.

When I arrive in L.A., I'll call your office and ask whether an appointment can be set up. If you wish to call me before then, you can reach me either by email or by phone, as noted above.

Sincerely,

Teresa Fernandez

Resume enclosed

See corresponding resume on page 255.

DENISE FRANCIS

6200 Skyline Blvd. • Oakland, CA 94611 • (510) 754-5151

Subject:
Cruise staff application

Dear Personnel Director,

Because of my combined interest in travel and working in a stimulating, unconventional environment with people of all kinds, I am applying for a cruise staff position with your firm.

My skills and personality are especially well suited to this work, as you'll note in the attached resume. I have years of successful experience working with many different types of people of all ages, and have sincerely enjoyed my work.

I would especially like to point out my strengths that are most relevant to the cruise field:
– strong in leadership and organizational skills;
– sensitive to the needs and feelings of others;
– personable, enthusiastic, and dependable.

I am 30, in fine health, and available for an extended cruise commitment.

I look forward to hearing from you at your earliest convenience. I can be reached at (510) 754-5151. (If I'm not home, please leave a message on my answering machine.)

Sincerely,

Denise Francis

See corresponding resume on page 35.

MARSHA RIFENBERG
12 Sherwood Avenue
Oakland, CA 94611
(510) 797-2131

Selection Committee
P.O. Box 4992-788
Walnut Creek, CA 94596

Re: Manager, Human Resources Development position

Dear Selection Committee:

In my current assignment in an engineering company, I derive my greatest satisfaction from providing management consultation and support. I enjoy working with management to explore new ways of motivating people, whether the issue is a job performance problem, policy/procedure interpretation or management/employee development. I also enjoy the results!

I have designed, delivered and evaluated management trainings. The focus of these trainings has been understanding and consistently applying company policies and procedures. I have also trained management in using various motivational techniques to inspire the best from their employees.

The enclosed resume describes my experience and skills in more detail.

I look forward to the opportunity to meet with you regarding your Manager, Human Resources Development position.

Sincerely,

Marsha Rifenberg

See corresponding resume on page 104.

JOHN BRIDGES
97 Foothill Lane
Berkeley, CA 94705
(510) 990-3466

Dr. Art Levinson
Chief Executive Officer
Genentech, Inc.
460 Point San Bruno Blvd.
South San Francisco, CA 94080

Dear Dr. Levinson,

It was a pleasure to attend the Genentech Shareholders Meeting last week.

After the meeting, I introduced myself to you and expressed my excitement at following the company as a shareholder, along with my desire to work directly for Genentech.

During the question period, I asked if the company had any plans for the treatment of breast cancer. The treatment of this disease and others through the activation and restoration of the immune system, with the immunoregulatory drugs created at Genentech, is of particular interest to me.

I would be thrilled by the opportunity to contribute to the work the company is doing in this field.

Enclosed is my resume which you kindly requested. Thank you very much for your interest and I look forward to hearing from you soon.

Sincerely,

John Bridges

See corresponding resume on page 184.

DONNA COLE
1776 - 12th Avenue
San Francisco, CA 94118
(415) 212-6822

August 12, 1996

Manager
Sales Department
P.O. Box 57346
Hayward, CA 94545

Dear Manager,

This letter is in response to your advertisement in Sunday's Examiner/ Chronicle seeking a salesperson. I was excited when I read your ad since I've had a long-time interest in food sales, and your product line sounds particularly appealing to me.

In the course of my recent career research, I have spent some time interviewing and accompanying a sales rep for a well known cookie and snack food manufacturer. That experience made me confident that I have the personality, aggressiveness, and persuasive manner required in this line of work.

I would be delighted to talk with you in person about this position, and look forward to hearing from you soon. I can be reached at 212-6822, where my answering machine will take your message if I'm out.

Sincerely,

Donna Cole

25 Tough Problems

and resumes that illustrate solutions

#1. Should I keep my resume **chronological** or switch to a **skills format?** 273

#2. What if I want a **more flexible resume**, not chronological OR functional? 273

#3. What if my **job objective is complicated** to describe? 273

#4. What if I have **more than one job objective**—or want a generic resume? 273

#5. How do I know **what skill areas** to put on my resume? 273

#6. What if I have **hardly any paid experience?** 274

#7. What can I do about **gaps in my work history?** 274

#8. What do I do with my **scrambled-up work record?** 274

#9. How do I get rid of the **"job-hopper" look** on my resume? 274

#10. What if my old **job title understated** my actual level of **responsibility?** 274

#11. What if I **don't have a degree,** but did take some classes? 275

#12. What if I **don't quite have my degree** or credentials yet? 275

#13. How do I show **part-time jobs while I was in school?** 275

#14. How can I avoid **age discrimination,** with a 30-year work history? 275

#15. What if I stayed with **one employer for a *very* long time?** 276

#16. What about **jobs I held many years ago?** 276

#17. What can I do about an **embarrassing job** in my work record? 276

#18. What if my last workplace has a **confusing-sounding name**? 276

#19. What about androgynous names that **aren't clearly male or female?** 276

#20. What's a good way to list **technical skills** or special knowledge? 276

#21. How could I include a more **personal statement** about how I work? 277

#22. How do I translate my **military background** to civilian? 277

#23. How do **foreign-born job hunters** list their education and work history? 277

#24. How do I adapt a resume to promote my **consulting services?** 277

#25. Can I make a **snazzy looking resume** with a standard word processor? 277

#1. Should I keep my resume chronological, or switch to a skill-based format?

General guideline: The standard **chronological** format works well if you're **staying in the same field** and moving up in your career—especially if your work history shows continuous progression, or if you've worked for prestigious companies.
A **skill-based** ("functional") resume may work better if you're **making a career change**—especially to a different field or to a much higher level of responsibility.

Look at the resumes below which illustrate the above guidelines on format choice:
Chronological resumes:

Page 19, Janet — continuous upward mobility in her chosen role of management
Page 154, Richard — seeks to move up in the emergency medical services field

Skill-based/functional resumes:

Page 242, Richard — applying his emergency-work skills to a new field, fire-fighting
Page 255, Denise — no previous cruise experience, but shows how her skills apply

#2. What if I want a more flexible resume format—not strictly chronological or strictly functional?

Look at the resumes below which combine the best features of functional resumes with the best features of chronological formats:

Page 44, Anne — accomplishments, then job content in chronological order
Page 54, Deborah — most recent job described from the skill perspective
Page 114, Charles — again, most recent job described in functional detail
Page 125, Michelle — two overlapping jobs, each broken down by skills

#3. What if my job objective is complicated to describe?

Look at the resumes below for ideas and solutions to this problem:

Page 56, Christie — strategic spacing and punctuation makes it easy to read
Page 58, Lani — Like Christie above, uses spacing and punctuation for clarity
Page 116, Christiane — makes it simpler and clearer through spacing and layout
Page 126, Sarah — bullets make it easy to see which classes she could teach
Page 228, Mary — a bold first line, and well-placed bullets, enhance readability
Page 258, Julia — two short columns, very easy to read

#4. What if I have several different job objectives—or I want a more generic resume I could use for all kinds of jobs?

General guideline: It is much more effective to use several different targeted resumes than one "generic" resume to cover a range of job possibilities.

*Here are examples of **two different targeted resumes** for the **same person**:*

Page 34, Carol — personnel analyst
Page 135, Carol — program development, elderly
Page 154, Richard — EMT Supervisor (EMT = Emergency Medical Technician)
Page 242, Richard — fire fighter, entry level

#5. How do I know what skill areas to put on my resume?

General guideline: The skills you need to document on your resume are those most **relevant to your stated job objective,** and you can find out what they are in several ways: by carefully reading the classified ad for the job, by reading the employer's official job description, or *(best of all)* by doing a little "informational interviewing," which is simply *talking directly with someone who already works in a similar job.*

The resume below was prepared after talking with someone in the field.

Page 232, Margaret — good results of informational interviewing

#6. What if I have hardly any paid experience?

Look at the resumes below for ideas and solutions to this problem:

Page 126, Sarah — expands fully on the experience she DOES have
Page 130, Hannah — volunteer experience given highest priority
page 42, Bill — taxi driver with volunteer experience
Page 201, Brian — lists all the coursework recently completed
page 219, Vicki — describes her class projects in detail
Page 224, Andrew — emphasizes his recent internship and volunteer work
Page 257, Matthew — does a lemonade stand count?

#7. What can I do about gaps in my work history?

General guideline: Make a POSITIVE, unapologetic statement about what you WERE doing.

Look at the resumes below for ideas and solutions to this problem:

Page 18, Molly — home management
Page 48, Stephen — grad student
Page 89, Claudia — maternity leave and family management
Page 99, Lydia — full-time parenting and graduate student
Page 152, Patricia — family management
Page 159, Carole — family/maternity leave
Page 164, Cynthia — avoids a gap while job-hunting
Page 175, Joanne — travel and full-time student

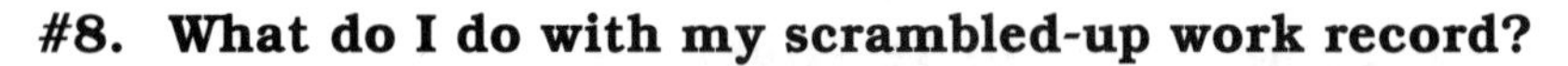

#8. What do I do with my scrambled-up work record?

Look at the resumes below for ideas and solutions to this problem:

Page 15, Helen — job progress is made clear
Page 22, Marla — explains the nature of the business where it's not obvious
Page 87, Mack — reverses the usual chronology to change the emphasis
Page 158, Carole — strategic formatting makes it clear
Page 167, Fran — consistency in layout makes for clarity
Page 190, Judy — a good compromise, where a lot of explaining is needed
Page 229, Mary — free-lance jobs presented in a compact paragraph
Page 230, Carolyn — creative layout makes it seem less confusing

#9. How do I get rid of the "job-hopper" look on my resume?

Look at the resumes below for ideas and solutions to this problem:

Page 72, Anthony — lumps temp jobs together
Page 81, Stephen — combines two short-term jobs
Page 87, Mack — odd jobs while seeking full-time employment

#10. What if my old job title understated my actual level of responsibility? Or didn't indicate what I really did?

If (for example) you were called "Administrative Assistant" or "Secretary" but in fact you had Office Manager responsibilities, then you could describe yourself fairly and **accurately** as "Office Manager."

(IMPORTANT NOTE: Remember, if the job was in the recent past and you intend to use your old boss as a reference, you will need to graciously let him know what you're doing and why, so he can back you up on this.)

Another example: In some major universities and government jobs, hundreds of people all have the same job title ("Senior Clerk," "Administrative Assistant," "Program Assistant"), which is actually a PAYROLL title. On the resume, it helps to add a job-descriptive title along side it. (**Student Academic Advisor**/Principal Clerk)

#11. What if I don't have a degree (or it isn't relevant) but I did take some classes?

Look at the resumes below for ideas and solutions to this problem:

Page 36, Margaret — mentions 112 college units
Page 46, Mary — shifts emphasis from her degree to her on-the-job training
Page 78, Joy — spells out her recent business training

#12. What if I don't quite have my degree or credentials yet?

In the EDUCATION section, you could say:

B.A. due April 1997, Accounting – Golden Gate University
or
M.A. candidate, Career Development JFK University (anticipated 6/97)

In the SUMMARY section, you could say:

- Graduate degree in Career Development pending at JFK University.
  or
- B.A. in Anthropology; elementary teaching credential pending.

Look at the resumes below for more ideas and solutions to this problem:

Page 94, Lynne — listing the academic major, where there's no degree
Page 117, Fereshteh — *eligible* for equivalent U.S. credentials
Page 165, Elizabeth — Graduate studies in expressive arts, in progress
Page 201, Brian — degree anticipated soon
Page 262, Rachael — B.A. in progress

#13. How do I show part-time jobs while I was in school?

Look at the resumes below for ideas and solutions to this problem:

Page 73, Anthony — shows that his work history is concurrent with schooling
Page 77, Estelle — uses "summer or "weekends" to explain
Page 81, Stephen — helped his father in the summer
Page 258, Julia — fourteen-year-old has lots of ambition
Page 259, Charlie — golf caddie and handyman
Page 260, Karen — summer internships
Page 261, Valerie — summer volunteer job

#14. How can I avoid age discrimination, with a 30-year work history?

General guideline: In both the Work History and Summary sections of your resume, ***you don't HAVE to mention ALL of your work history.*** You can go back only as far as necessary (say 10–15 years) to document a substantial—but not TOO substantial—record of employment.

For example, *even though you may actually have 30 years of experience* as a teacher, it is ALSO honest—and may serve you better—to say in the SUMMARY section:

- Fifteen years experience in public school teaching.
  OR
- Over 15 years experience in public school teaching.

This tactic has been used on a number of the resumes in this book.

Eventually (at the job interview) you may have to deal with age discrimination, but at least you'll have a chance to get your foot in the door FIRST.

#15. What if I stayed with one employer for a very long time?

Look at the resumes below for ideas on how to present lengthy experience:

Page 15, Helen — shows advancement in her field while staying with one employer
Page 90, David — lists each position at one place, showing range of job titles
Page 172, Betsy — shows variety and responsibility
Page 230, Carolyn — focuses on advancement in the job

#16. What about outdated experience from jobs I held many years ago?

Look at the resumes below for ideas and solutions to this problem:

Page 113, Tudy — prior years: variety of positions
Page 197, Vreny — plus earlier store management in Switzerland

#17. What can I do about an embarrassing job in my work record?

General guideline:

Don't put things on your resume that could undermine your desired image.

For example,

Dolores was embarrassed at the name of her old workplace, "Uncle Bunnie's Incredible Edible," considering her relatively dignified current objective. But the experience was too important to omit, so we took liberties and *changed the name of the workplace* to "U.B.I.E. Restaurant," like so:

1985 ***Night Manager*** U.B.I.E. RESTAURANT, Rockridge CA

Another example:

Page 187, Jerry — "Truck Driver" changed to "Transportation."

#18. What if my last workplace has a confusing name?

The names of some businesses give **no clue to what they are about**, so merely listing them in your Work History doesn't give the reader enough information.

Look at the resumes below for ideas and solutions to this problem:

Page 29, Rita — Seventy-Seven, Inc. = Hawaii real estate development
Page 68, Gelia — Shoreline Shipping, Inc. = international cargo shippers
Page 205, Rebecca — Bay City Unique Creations = private label design firm

#19. What about androgynous names that aren't clearly male or female?

General guideline: Don't leave people in doubt. Even though it may be irrelevant to the job, readers are uneasy when they can't tell from your name whether you are a man or a woman.

Adding your middle name, or a Mr. or Ms. prefix, could solve this problem. See:

Page 186, Hollis — adds her middle name, "Ann"
Page 95, YonSoon — adds "Ms." prefix

#20. What's a good way to list technical skills or special knowledge?

Look at the resumes below for layout ideas:

Page 72, Anthony — coursework
Page 87, Mack — accounting coursework
Page 104, John — lab skills
Page 201, Brian — coursework and equipment
Page 209, Martha — video/film equipment
Page 232, Margaret — home maintenance tools and skills

#21. How could I include a more personal statement about how I work?

Look at the resumes below for ways to include a personal touch on your resume:

Page 42, Bill — tells what he believes in
Page 151, Ken — describes his professional approach and techniques
Page 236, Sister Mary — makes clear what motivates her

#22. How do I translate my military background to civilian?

Look at the resumes below for ideas and solutions to this problem:

Page 8, Charmaine — Army management experience applied to nonprofits
Page 46, Mary — emphasizing management training in the Army

#23. How do foreign-born job hunters list their education and work history?

See the resumes below for ideas on listing education and employment outside the U.S.

Page 16, Hellmut — Germany
Page 86, Sonia — Guatemala
Page 95, YonSoon — Korea
Page 103, Mark — Iran
Page 116, Christiane — Germany
Page 117, Fereshteh — Iran
Page 118, Sharon — France
Page 197, Vreny — Switzerland

#24. How can I adapt my resume so it's clear that I'm promoting my professional or CONSULTING services—not looking for a job?

These people are seeking a professional affiliation, rather than employee status:

Page 12, Diane — health care management
Page 100, Rochelle — PC support specialist
Page 110, Nicholas — educational program development
Page 120, Elizabeth — educational program development
Page 151, Ken — body therapist
Page 152, Patricia — stress reduction and biofeedback
Page 158, Carole — medical / nursing
Page 228, Mary — free-lance writer/editor
Page 246, Donald — private investigator

#25. How can I make a snazzy looking resume when I only have a standard word processor?

Look at the resumes below for ideas on attractive designs requiring only Microsoft Word and one or two fonts:

Page 15, Helen — horizontal lines and dingbats add appeal
Page 50, Christine — gray-shaded decorative bars for elegance
Page 68, Gelia — another variation of line-and-dingbat design
Page 106, Patricia — two fonts (Souvenir and Zapf Dingbats) plus a little shading
Page 161, Tracy — a design from "Ready-to-Go Resumes" software by Yana
Page 183, Denise — another "Ready-to-Go Resumes" design
Page 185, Ellen — a variation on the gray-shaded bar design
Page 218, Lynne — another elegantly simple design by Ron Weisberg

JOB OBJECTIVES

Account executive, 170, 191
Accountant, 83, 84, 86, 89, 95
Activity director, 255
Addictions counselor, 130
Administrative analyst, 32, 38, 48
Administrative assistant, 78
Administrator, 10, 54, 66
Adult education teacher, 126
Art teacher, 118
Attorney, 220
Auditor, 88
Auto manufacturer's service writer, 244
Baby-sitting, 258
Baker's apprentice, 251
Benefits advocate, 42
Biofeedback specialist 152
Biotechnology research, 104
Body therapy, 151
Bookkeeping, 77, 91, 92
Bookstore clerk, 74
Business development, 79
Business manager, 94
Buying, 186
Career counseling, 113, 125
Chemical dependency treatment, 136, 140
Child care, 117
Client services, 173, 178, 206
Clothing design/manufacturing, 202, 204, 205
College admissions officer, 66
Community development, 48
Computer consultant, 100
Computer support services, 100, 106
Construction project supervisor, 50
Consultant, 12, 84, 93, 120
Copyediting, 228
Counselor, addictions, 130
Counselor, family, 132
Counselor, mental health, 124, 133
Counselor, school guidance, 114
Cruise staff, 255
Customer service, 64, 72, 76, 244
Data management, 101
Department manager, 197
Director, 16, 18, 22, 28
Display/merchandise, 188
Editor, 228
Editorial assistant, 226, 227, 230
Educational program development, 110, 120
Emergency medical services, 154
Engineer/plastics, 98
Engineering management, 105
Entry-level objectives, 81, 83, 87, 88, 174, 211, 212, 217, 233, 242, 251, 259-262
Environmental resource manager, 200
Environmental technician, 201
Events coordinator, 57, 62, 177, 255
Executive assistant, 29, 79
Executive director, 28
Fashion, 177, 195
Film production, 207, 210
Financial analyst, 90, 94
Financial consultant, 84, 93
Financial planning, 94
Fire fighter, 242
Food and drug quality control, 7
Franchise management, 30
Freight handler, 253
Fund-raiser, 38, 62
Gallery assistant, 252
Governmental relations rep, 241
Graphic designer, 215, 217
Graphic production, 216
Health care management, 9, 12, 14
Health care services, 70, 156, 178
Health educator, 146
Health services administration, 14, 32
Hotel sales, 182
Housing rehabilitation, 24
Human resources admin, 35, 40
Import/export, 253
Information services, 52, 200
Instructional design, 106
International business, 29, 65, 79, 253
Internet technology, 106
Inventory control, 68
Investigator, 246
Jewelry sales, 193
Labor organizer, 238
Language instructor, 116
Legal assistant, 222, 223, 224, 240, 262
Legal summarizer, 240
Library assistant, 112
Management services, 26
Management/manager, 7, 8, 9, 12, 14, 18, 20, 22, 26, 30, 44, 94, 144, 179
Manufacturer's rep, 185, 244
Manufacturing support, textiles, 58
Marketing, 162, 163, 164, 166, 168, 171, 174, 175, 179, 181, 190, 196, 260
Market research, 161
Massage, 151
Medical consultant, 12, 152, 158
Medical emergency services, 154
Medical investigator, 143
Medical products, 179, 184
Medical services supervisor, 154
Member relations, 70
Mental health services, 124, 133
Merchandising display, 188
Minister, 234
Multi-media design, 219
Nonprofit administration, 8, 10, 16, 18, 22, 56, 62
Nursing administration, 15, 156
Nutritionist, 146
Odd jobs, 258, 259
Office assistance/support, 76, 77, 80, 81, 261
Office management, 19
Online technologies instructor, 106
Organizational development, 10, 46
Paralegal, 223
Pasteup artist, 215
Pastor, 234, 236
Personnel officer, 34, 35, 78
Photographer, 239
Physical therapy aide, 157
Plastics engineer, 98
Private investigator, 246
Product design, 202, 206
Production assistant, 205
Program development for elderly, 135
Program/project coordination, 32, 56, 59, 75, 78, 148, 206
Program/project director, 37, 50, 59, 60, 69, 172, 173
Property management, 212, 213, 214
Psychotherapist, 134, 140
Public information services, 52
Public relations, 56, 64, 162, 165, 166, 170, 172, 176, 196
Publications designer, 218
Purchasing, 50, 68
Quality control, 7
Radio announcer, 208
Radiology manager, 108
Real estate analyst, 245
Real estate appraiser, 231, 232, 233
Recruiter, 36
Research services, 101, 104, 200
Resources management, 200
Sales management, 168, 171
Sales rep, 183, 187, 192, 195, 198
Sales, 171, 181, 184, 186, 190, 191, 193, 194, 195, 196, 254
School guidance counselor, 114
Security consultant, 246
Social service administration, 16, 69
Social worker, 131, 132, 135, 138
Software support, 89, 97
Store manager, 197
Stress reduction specialist, 152
Teacher, 118, 124, 126, 128
Technical writer, 102
Telecommunications supervisor, 103
Therapist, 132, 133, 134, 140
Tour escort, 255
Training specialist, 33, 106
Translator, 122
TV news grip, 211
TV programming, 208, 209
Union rep, 238
Volunteer coordinator, 67
Wardrobe assistant, 248
Warehouseman, 250
Workshop presenter, 125
Writer, 102, 228, 239

INDEX

NOTE: **The five main features of a resume** — Job objective, Skills, Experience, Education, Work History, are in **BOLD** for quick reference.

Age discrimination, 275
Androgynous names, 276
Attractive resume design, 277
Children's resumes, 257, 258
Chronological format,
  when to use, 273
College students/grads. *See* Students
Consulting, resumes for, 277
Cover letters, 263-271
Credentials, incomplete, 275
Damn Good Resume
  Ten Steps to, 281
  vs. perfect resume, 1
Degree, lack of, 275
Education, incomplete, 275
Electronic Resume Revolution, 3
Embarrassing job title, 276
Employment history. *See* Work history
Experience
  lack of, 274
  outdated, 276
Foreign-born, 277
Format
  attractive design, 277
  choosing, 273
  chronological, 273
  combined/flexible, 273
  functional, 273
Freelancing, resumes for, 277
Functional format, when to use, 273
Gaps, in work history, 273
Generic resume, 274
Job history. *See* Work history
Job-hopper image, 274
Job objective(s)
  complex, 273
  freelancing/consulting, 277
  index of, 278
  multiple, 273
Job title
  embarrassing, 276
  inadequate, 274
Kennedy, Joyce Lain, 3
Key words, in scannable resume, 3
Military background, 277
Multiple resumes, 116, 253, 273
Part-time jobs, listing them, 275
Personal philosophy, stating, 277
Problems on resumes, 272-277
  age discrimination, 275
  androgynous names, 276
  with education/credentials:
    coursework only, 275
    incomplete credentials, 275
    overqualified, 102, 208
  foreign-born, 277
  format, confusion in choosing, 273
Problems on resumes (continued)
  freelancing, 277
  identifying relevant skill areas, 273
  in work history. *See* Work History Problems
  job objective(s)
    complex, 273
    multiple, 273
  listing technical skills, 276
  military background, 277
  personal philosophy, stating, 277
Resume, "damn good" vs. "perfect," 1
Resume format. *See* Format
Resume design, attractive, 277
Resume Pro newsletter, 282
Resume Pro: the Professional's Guide, 280
Resume samples, listed, 4-5
Resume samples, 6-262
  Administration/coordination, 49
  Education, 109
  Finance/accounting, 82
  Health, 142
  Human resources, 31
  Management, 6
  Marketing/promotion, 160
  Office/administrative support, 71
  Potpourri, 199
  Sales, 180
  Technical and computers, 96
  Therapy and social work, 129
  Youth and Students, 256
Scanner-friendly resume, 2
Skill areas, which to include, 273
Students and recent grads,
  college address, 260
  coursework but no degree, 274
  incomplete degree, 275
  listing summer jobs, 275
  sample resumes 72, 77, 81, 219, 257-262,
Teaching aids (resume writing), 281
Technical skills, how to list, 276
Ten Steps to a Damn Good Resume, 281
Work experience abroad, 124, 128, 277
Work history problems
  experience
    lack of, 274
    outdated, 276
  gaps in employment, 273
  job-hopper image, 274
  job title
    embarrassing, 276
    inadequate, 274
  listing part-time/summer jobs, 275
  scrambled work record, 274
  too long with one employer, 276
  workplace name confusing, 276
Youth resumes, 257, 258

Yana Parker's

Resume Trainer's Kit

The Resume Trainer's Kit is a collection of high quality workshop or classroom materials for effectively teaching resume writing to various client populations.

Basic text, workshop outlines, exercises, handouts—are all in one package. YOUR Trainer's Kit can be customized with worksheets designed for your specific population. A Kit may contain any or all of the following instructional materials:

- **Fill-in-the-blanks RESUME WORKSHEETS** to guide job hunters or students in preparing a hand-written first draft of their resume. The filled-in worksheets can be used in various ways:
  - ➢ by agency staff, as an efficient guide in typing up a final resume;
  - ➢ by the counselor, to critique the client's work-in-progress and advise on improvements.
  - ➢ by the job-hunter, as a first step in producing a final word-processed resume.

  These WORKSHEETS have been very carefully designed for maximum effectiveness, in both functional and chronological formats.

  Note: Customized versions will be available for high-school students, job-hunters leaving the military, job-hunters transitioning from public assistance, ex-offenders returning to the workforce, displaced homemakers / parents.

- **Fill-in-the-blanks WORKSHEETS for related documents:**
  - ➢ Worksheet / Guide for Writing an Effective Recommendation Letter
  - ➢ Worksheet / Guide for a Verbal Reference
  - ➢ Worksheet / Guide for Writing a Cover Letter
  - ➢ Worksheet / Guide for Writing a Follow-up or Thank You Letter

- **HANDOUTS** (hard copy and/or overheads) that graphically depict key concepts, formats, and designs, and provide illustrative examples and strategies (some customized for different populations):
  - ➢ Workshop Outline(s)
  - ➢ Workshop Evaluation Form
  - ➢ Exercises for identifying transferable skills
  - ➢ Exercises for identifying appropriate job goals
  - ➢ Exercises for identifying marketable experience and accomplishments
  - ➢ Sample resumes, chronological and functional
  - ➢ Guide to Selecting a Resume Format
  - ➢ Resume Templates (functional / chronological; plain / fancy; simple / sophisticated)
  - ➢ Guide for an Upgrade (promotional) Resume
  - ➢ Guide for Creating a Scanner-Friendly Resume
  - ➢ Introduction to Resume Banks on the Internet
  - ➢ Allocating Space on Your Resume
  - ➢ Guide to Informational Interviewing
  - ➢ Examples of transferable skill sets for various job objectives
  - ➢ Examples of resume techniques that resolve problems
  - ➢ Examples of good accomplishment statements
  - ➢ Examples of good summary / profile statements
  - ➢ Hot Tips on Resume Writing
  - ➢ Sample Make-Over Resumes (before / after)

- **Poster** of the TEN STEPS to a Damn Good Resume ("Darn" version MAY become available!) The poster is 24" × 36" classroom size, in color.

- **Bookmark Mini-Guides:** 3" × 8" card-stock guides repeating the same Ten Steps to a Great Resume that appear in Damn Good Resume Guide and on the Poster described above.

- **Books** • *Damn Good Resume Guide* • *Resume Catalog: 200 Damn Good Examples* • *Blue Collar & Beyond: Resumes for Skilled Trades and Services* • *Ready-To-Go Resumes* software / templates.

WRITE, FAX, or CALL Yana for current PRICE and CONFIGURATION of the Trainer's Kit.

Poster: 24 x 36 inches

Ten Steps to a Great Resume

1 **Choose a job target** (also called a **job objective**). An actual job title works best.

2 **Find out what skills, knowledge, and experience are neede** to do that target job.

3 **Make a list of your 3 or 4 strongest skills or abilities or knowledge** that make you a good candidate for the target

4 **For each key skill, think of several accomplishments** fr past work history that illustrate that skill.

5 **Describe each accomplishment** in a simple, powerfu **statement** that emphasizes the results that benefite employer.

6 **Make a list of the primary jobs you've held, in ch order.** Include any unpaid work that fills a gap or you have the skills for the job.

7 **Make a list of your training and educational** are related to the new job you want.

8 **Choose a resume format that fits your situ** chronological or functional. (Functional wor changing job fields; chronological works b up in the same field.)

9 **Arrange your action statements** accor you chose.

10 **Summarize your key points** at or nea

(In real-life resume writing, we do skip around. together in some other sequence—as long a

From *The Damn Good Resume Guide*, 3rd edition, © 1996, 1989, 1986, 1983 by Yana Parker. Published by Ten Speed Press, P.O. Box

New from Yana Parker!

Bookmark: 3½ x 8½ inches

– A Bookmark Mini-Guide –

TEN STEPS to a Damn Good Resume

1. **Choose a job target** (also called a "job objective"). An actual **job title** works best.
2. **Find out what skills, knowledge, and experience are needed** to do that target job.
3. **Make a list of your 3 or 4 strongest skills or abilities or knowledge** that make you a good candidate for the target job.
4. **For each key skill, think of several accomplishments** from your past work history that illustrate that skill.
5. **Describe each accomplishment** in a simple, powerful, action statement that emphasizes the results that benefited your employer.
6. **Make a list of the primary jobs you've held, in chronological order.** Include any unpaid work that fills a gap or that shows you have the skills for the job.
7. **Make a list of your training and education** that are related to the new job you want.
8. **Choose a resume format that fits your situation**—either chronological or functional. (Functional works best if you're changing fields; chronological works best if you're moving up in the same field.)
9. **Arrange your action statements** according to the format you chose.
10. **Summarize your key points** at or near the top of your resume.

In REAL-LIFE resume writing, we DO skip around. So don't worry if YOUR resume comes together in some other sequence — as long as you do #1 and #2 first!

— Check out the other side of this bookmark, too. —

This much-requested advice, gathered from Yana's five popular books, is now available in two fomats: an attractive 24-by-36 inch poster and a handy bookmark (the bookmark also includes **Five Key Concepts For a Powerful, Effective Resume.**) Great for teachers, career counselors, and others, these reasonably priced items are available from the publisher. For price and ordering information, contact:

The Order Department
Ten Speed Press
P.O. Box 7123
Berkeley, CA 94707
(800) 841-BOOK

"Damn Good" Resume Pro Newsletter

A national newsletter for professionals, exploring and promoting excellence in resume writing

A quarterly publication since 1980 ◆ Editor: Yana Parker

How to Subscribe

Resume Pro Newsletter is a quarterly 20-page publication for professional resume writers, career counselors, human resource staff, educators, entrepreneurs ... all about excellence in resume writing.

Each issue includes:

- ◆ More "damn good" resume examples in a wide range of career fields. Before-and-after "make-overs" illustrating the principles of powerful resume writing and effective graphic design for resumes.
- ◆ Feature articles on great resume writing and effective job hunting.
- ◆ Problem-solving tips; creative ways to write a good resume despite a less-than-perfect background.
- ◆ Letters to and from counselors, writers, and educators, on innovative programs around the country that serve job hunters.
- ◆ Tips for getting maximum benefit from emergng technolgies such as electronic scanning and vast internet resources.

"Damn Good" Resume Pro Newsletter

A national newsletter for professionals, exploring and promoting excellence in resume writing

Summer 1996 • Editor: Yana Parker • Issue #25

AT A RESUME WORKSHOP

MY BASIC PHILOSOPHY: On Resumes and Work

In This Issue

The Damn Good Resume Pro Newsletter

A national newsletter for professionals, exploring and promoting excellence in resume writing

Fall 1993 • Issue #14 • Publisher, Yana Parker

Making Friends with Resume Scanning: Success Strategies for Resume Writers

In This Issue

- Resume Scanning - 1
- Resume Clinic - 2
- Gilda Weinskopf's Business - 3
- Ask Clara about Scanning - 5
- Pow!Litzer Prize Winner - 5-6
- Accomplishments - 7
- Resume Formats - 8-9
- Resume Design - 10-12
- To the Editor - 13
- Volunteering Pays Off - 14
- Peeking into Offices - 15
- Volunteers Wanted - 15
- Subscriptions - 15
- This Year's Workshop - 16

To subscribe ...

Check or circle what you want:

❑ $25 for one year (4 issues) within the U.S.A..

❑ $30 (US dollars) for one year (4 issues) to Canada and other countries.

Please make your check payable to "Yana Parker" and send to: PO Box 3289, Berkeley CA 94703.

Date ______________________

Name ______________________

Company/Title ______________________

Street ______________________

City/State/Zip ______________________

Work phone ____________ Home ____________

Fax phone ______________________

More Damn Good Advice from Yana

THE DAMN GOOD RESUME GUIDE

The new edition of this classic best-seller, with over half a million copies in print, now features a failproof step-by-step approach that job-hunters of any age or level can easily follow to produce a *damn good* resume. At each step along the way, users are deftly guided past potential stumbling blocks with creative solutions to tough problems such as lack of experience and gaps in one's work history. Warmth and humor couples with savvy down-to-earth advice to keep this perennial favorite on the must-have list for career centers and job hunters across the county. 80 pages.

BLUE COLLAR AND BEYOND:
Resumes for Skilled Trades and Services

What we once called "blue collar" jobs have expanded to include all kinds of work—not just auto repair and construction, but food service, hospitality, clerical support, health and beauty, and so on. These jobs often require resumes, but most resume books don't address them. Yana does, giving the basics of resume writing—including how to deal with difficult or embarrassing issues—and provides well over 100 sample resumes to show how it's done. 192 pages

READY•TO•GO RESUMES
Self-Teaching Resume Templates for Mac and PC

Based on Yana's best-selling resume books (including the one you're reading right now), this book/software package contains all the tools that any computer-using job-hunter will need to craft a damn good resume. The helpful book gives resume-writing basics, then walks the reader through the process of creating a tailor-made resume, using the enclosed software designed by Yana. The book is packaged with three disks, each holding 28 resume templates—that's one disk each for Word for Mac, Word for Windows, and WordPerfect for Windows. 128 pages

RESUME PRO:
Make Money Writing Resumes

Once you've learned the resume basics in this book, it may occur to you that resume-writing is a lot more fun than job-hunting. This book tells you how to set up and operate your own resume-writing service. It is equally helpful for career counselors and managers who must write a lot of very different types of resumes for their clients. Packed with trade secrets from one of the best pros in the field. 416 pages

For more information, or to order, call the publisher at the number below. We accept VISA, MasterCard, and American Express. You may also wish to write for our free catalog of over 500 books, posters, and audiotapes.

Ten Speed Press
P.O. Box 7123
Berkeley, CA 94707
(800) 841-BOOK